Springer Theses

Recognizing Outstanding Ph.D. Research

Aims and Scope

The series "Springer Theses" brings together a selection of the very best Ph.D. theses from around the world and across the physical sciences. Nominated and endorsed by two recognized specialists, each published volume has been selected for its scientific excellence and the high impact of its contents for the pertinent field of research. For greater accessibility to non-specialists, the published versions include an extended introduction, as well as a foreword by the student's supervisor explaining the special relevance of the work for the field. As a whole, the series will provide a valuable resource both for newcomers to the research fields described, and for other scientists seeking detailed background information on special questions. Finally, it provides an accredited documentation of the valuable contributions made by today's younger generation of scientists.

Theses are accepted into the series by invited nomination only and must fulfill all of the following criteria

- They must be written in good English.
- The topic should fall within the confines of Chemistry, Physics, Earth Sciences, Engineering and related interdisciplinary fields such as Materials, Nanoscience, Chemical Engineering, Complex Systems and Biophysics.
- The work reported in the thesis must represent a significant scientific advance.
- If the thesis includes previously published material, permission to reproduce this must be gained from the respective copyright holder.
- They must have been examined and passed during the 12 months prior to nomination.
- Each thesis should include a foreword by the supervisor outlining the significance of its content.
- The theses should have a clearly defined structure including an introduction accessible to scientists not expert in that particular field.

More information about this series at http://www.springer.com/series/8790

Nurul T. Islam

High-Rate, High-Dimensional Quantum Key Distribution Systems

Doctoral Thesis accepted by Duke University, Durham, NC, USA

 Springer

Nurul T. Islam
Department of Physics
Duke University
Durham, NC, USA

ISSN 2190-5053   ISSN 2190-5061   (electronic)
Springer Theses
ISBN 978-3-030-07548-4   ISBN 978-3-319-98929-7   (eBook)
https://doi.org/10.1007/978-3-319-98929-7

This Springer imprint is published by the registered company Springer Nature Switzerland AG.
The registered company address is: Gewerbestrasse 11, 6330 Cham, Switzerland

This dissertation is dedicated to my parents.

Supervisor's Foreword

Dr. Nurul T. Islam's thesis describes a broad research program on the topic of quantum communication. Here, a cryptographic key is exchanged by two parties using quantum states of light and the security of the system arises from the fundamental properties of quantum mechanics. Dr. Islam develops new communication protocols using high-dimensional quantum states so that more than one classical bit is transferred by each photon. This approach helps circumvent some of the non-ideal properties of the experimental system, allowing him to achieve record key rates on metropolitan distance scales. Another important aspect of the work is the encoding of the key on high-dimensional phase-randomized weak coherent states, combined with the so-called decoy states to thwart a class of possible attacks on the system. The experiments are backed up by a rigorous security analysis of the system, which accounts for all known device non-idealities. Dr. Islam goes on to demonstrate a scalable approach for increasing the dimension of the quantum states and considers attacks on the system that use optimal quantum cloning techniques. The thesis captures the current state of the art of the field of quantum communication in laboratory systems and demonstrates that phase-randomized weak coherent states have application beyond quantum communication.

In greater detail, the introductory Chap. 1 introduces the basics of cryptography and points to the need for post-quantum cryptographic methods that will be secure in the presence of an attacker with a large-scale quantum computer. Quantum key distribution offers one potential solution, but additional research is needed to improve the secure key rate of these systems and to analyze their security in the presence of experimental imperfections. The chapter then introduces the dissertation, highlighting the original contributions to the field. Chapter 2 is a pedagogical introduction to quantum key distribution using two-dimensional (qubit) time-phase states. Here, a quantum photonic wavepacket is in one of two contingent time slots in the communication system (temporal basis) or in both time slots with a relative phase difference (phase states). The reader is walked through a quantum key distribution session using these states and some possible attacks by an eavesdropper are discussed. The chapter is rounded out by discussing a possible experimental realization of a quantum key distribution system, highlighting how the phase states can be measured at the receiver using thermally insensitive delay interferometers (described in greater detail in Appendices A and B). A thorough discussion of a four-dimensional time-phase quantum key distribution system is presented in Chap. 3. Here, Dr. Islam used phase-randomized weak coherent states at the transmitter and identifies the fraction of the transmitted states for which the photonic wavepacket has one or less photons using the decoy-state protocol. A current record-high secure rate is obtained with this system for channel losses useful for a metropolitan-scale quantum key distribution network. Dr. Islam achieved this result by paying strict attention to experimental non-ideal behaviors, especially single-photon detector saturation. The experimental work is complemented with a full security analysis using entropic uncertainty relations, a state-of-the-art theoretical approach for this problem with additional details given in Appendix C. Following in Chap. 4 is a security analysis for a quantum key distribution system does not use all quantum states. A quantum key distribution system is secured using at least two mutually unbiased bases. For a d-dimension system, there are d states in each basis and each state on both bases needs to be chosen randomly and sent by transmitter. An analysis is presented here to show that only d states in one basis and one state in the other basis need to be used without substantial loss in key rate provided that the quantum bit error rate is low. This result is important for practical application for high-dimensional quantum key distribution systems and can be applied to any quantum photonic states. Going beyond a four-dimensional Hilbert space used in Chap. 3, Dr. Islam presents in Chap. 5 a new method for measuring phase states using a weak local oscillator and two-photon interference at the receiver. This approach does not require delay interferometers and hence is readily scaled to higher dimensions. Using this system along with the efficient security approach described in Chap. 4, Dr. Islam observes that detector saturation reduces the key rate at low channel loss; increasing the dimension helps avoid saturation effects, but eventually lowers the rate for large dimension. Thus, the secure key rate is optimized for a dimension that is always larger than two for this system. Chapter 6 describes one type of attack on a high-dimensional quantum key distribution system enabled by optimal quantum cloning. Previous experiments on optimal quantum cloning use heralded single-photon states generated by parametric down conversion. Here, Dr. Islam demonstrates that optimal quantum cloning is also possible using phase-randomized weak coherent states with the decoy-state protocol, which is well matched to the transmitter used in the quantum key distribution system. Consistent with previous work, the fidelity of the cloned state decreases with dimension and hence the security of the system increases assuming constant bit error rate. Finally, Dr. Islam summarizes his original contributions and points to future directions in the final conclusion chapter.

Columbus, OH   Daniel J. Gauthier

Acknowledgments

I would like to acknowledge the guidance and support of my adviser, Prof. Daniel Gauthier. Over the last 4 years, I have greatly appreciated and benefited from the amount of time he dedicates to his students, the way he motivates and encourages everyone to collect better data, and the rate at which he responds to email queries. It is an absolute privilege to learn under his guidance. I would like to thank Dan for putting up with me over the last 4 years, and his willingness to continue to do so.

I would also like to acknowledge the support of Prof. Jungsang Kim, who after Dan moved to OSU, provided me with a lab space, equipment, and an incredible environment to work. I would also like to thank him for realizing that the high-efficiency single-photon detectors were critical for these projects. On the same note, I would also like to thank Clinton Cahall for his work on the cryogenic readout circuits that I have used to collect most of the data presented here. Finally, I would like to thank Dr. Charles Ci Wen Lim for teaching me the basics of security proofs and for patiently explaining to me why I should always check my intuition with calculation.

I am grateful to my preliminary and thesis committee members, Prof. Harold Baranger, Prof. Kate Scholberg, and Prof. Henry Everitt, for the helpful discussion and questions and for taking the time to read through this work.

I would like to acknowledge all the past members of the Qelectron research group who overlapped a portion of their graduate school or postdoctoral life with me—Michael, Bonnie, Hannah, Andres, Otti, Lou, David, Nick—and the only other present member Meg, for all the conversations, laughter, physics-related discussions, etc.

During graduate school, it is very easy to lose perspective on the bigger picture of life. I would like to thank Arya Roy for constantly reminding me of that, and all the free coffee and great conversations about history, politics, sports, and physics. For the same reasons, a very special thanks to my friends of grads13—Gleb, Yingru, Agheal, Ming—who kept me going with frequent lunch meetings and hangouts. I would like to thank Payal for the company during long work days and evenings and for proofreading most of this thesis.

I also want to thank my college Professors, Dr. Harmon and Dr. Haring-Kaye, for constantly checking-in with me during my time at Duke. I want to thank Mrs. Harmon for all the free food and care. Finally, I would like to thank my parents and sister for their constant support, understanding, and patience. They have always managed to keep me shielded from the day-to-day activities at home.

I would like to gratefully acknowledge the support of the Office of Naval Research MURI program on Wavelength-Agile Quantum Key Distribution in a Marine Environment, Grant # N00014-13-1-0627, for funding my research work at Duke.

Contents

Parts of this thesis have been published in the following journal articles:

- Andres Aragoneses, Nurul T. Islam, Michael Eggleston, Arturo Lezama, Jungsang Kim, and Daniel J. Gauthier, *Bounding the outcome of a two-photon interference measurement using weak coherent states*, arXiv:1806.05012 (2018).
- Nurul T. Islam, Clinton Cahall, Charles C. W. Lim, Jungsang Kim, Daniel J. Gauthier, *Securing quantum key distribution systems using fewer states*, Phys. Rev. A, **97**, 042347 (2018).
- Joseph Lukens, Nurul T. Islam, Charles C. W. Lim, Daniel J. Gauthier, *Reconfigurable, single-mode mutually unbiased bases for time-bin qudits*, Appl. Phys, Lett., **112**, 111102 (2018).
- Clinton Cahall, Kathryn L. Nicolich, Nurul T. Islam, Gregory P. Lafayatis, Aaron J. Miller, Daniel J. Gauthier, Jungsang Kim, *Multi-Photon Detection using a Conventional Superconducting Nanowire Single-Photon Detector*, Optica **4**, 1534 (2017).
- Nurul T. Islam, Charles C. W. Lim, Clinton Cahall, Jungsang Kim, Daniel J. Gauthier, *Provably-Secure and High-Rate Quantum Key Distribution with Time-Bin Qudits*, Sci. Adv. **3**, e1701491 (2017).
- Nurul T. Islam, Andres Aragoneses, Arturo Lezama, Jungsang Kim, Daniel J. Gauthier, *Robust and stable delay interferometers with application to d-dimensional time-frequency quantum key distribution*, Phys. Rev. Appl. **7**, 044010 (2017).
- Nurul Taimur Islam, Clinton Cahall, Andres Aragoneses, Charles Ci Wen Lim, Michael S. Allman, Varun Verma, Sae Woo Nam, Jungsang Kim, Daniel J. Gauthier, *Enhancing the secure key rate in a quantum-key-distribution system using discrete-variable, high-dimensional, time-frequency states*, Proc. SPIE, Quantum Inf. Sci. Technol. II 9996, 99960C (2016). http://dx.doi.org/10.1117/12.2241429

List of Abbreviations and Symbols

Symbols

ℓ	Secret key length
$\lvert t_n \rangle$	Time basis states
$\lvert f_n \rangle$	Phase basis states
$H(x)$	d-Dimensional Shannon entropy for probability x

Abbreviations

cw	Continuous wave
DI	Delay interferometer
HOM	Hong-Ou-Mandel
IM	Intensity modulator
MUB	Mutually unbiased basis
PM	Phase modulator
PRWCS	Phase randomized weak coherent states
SNSPD	Superconducting nano-wire single-photon detector
UQCM	Universal quantum cloning machine

Chapter 1
Introduction

Development of scalable quantum computing platforms is one of the rapidly expanding areas of research in quantum information science [1, 2]. With many companies working towards building these platforms, a medium-scale quantum computer capable of demonstrating quantum supremacy over classical computers is in earnest only a few years away. Quantum computers pose a serious threat to the cybersecurity because most of the current cryptosystems, like the one devised by Rivest, Shamir and Adleman (known as the RSA), are based on computational hardness assumptions. For example, in an RSA protocol, to decrypt the private key an eavesdropper (Eve) needs to factorize a large number into its prime factors, which is believed to be a difficult problem for a classical computer. The most efficient classical algorithm that is known to factorize large numbers requires an exponentially large number of operations, which makes it difficult to solve the problem in practical timescales. Therefore, the security of such a protocol relies on the limitation of an eavesdropper's power and resources.

In a seminal paper published in 1994, Peter Shor showed that a powerful eavesdropper with access to a quantum computer can potentially solve the factorization problem with polynomial number of operations and in practical timescales [3–5]. Hence, cryptosystems in the forthcoming era of quantum computers need to be quantum-safe. In other words, a cryptosystem must be able to transmit a secret key even when an eavesdropper has access to quantum computers.

Quantum key distribution (QKD) with symmetric encryption is one of the very few methods that can provide provable security against an attack aided with a quantum computer [6]. The first QKD scheme was proposed by Charles Bennett and Giles Brassard in 1984, which later became known as BB84 [7]. Since its inception, QKD has often been featured as a secure communication technique whose security is guaranteed by the fundamental properties of quantum mechanics. The so-called "unconditional" security of a QKD system comes from two observations in

© Springer Nature Switzerland AG 2018
N. T. Islam, *High-Rate, High-Dimensional Quantum Key Distribution Systems*,
Springer Theses, https://doi.org/10.1007/978-3-319-98929-7_1

quantum mechanics. First, any eavesdropper trying to gain information about the transmitted bits from a sender (Alice) will inadvertently disturb the fidelity of the quantum states, which can be detected at the receiver (Bob). Second, Eve cannot duplicate any unknown quantum state, even with access to a quantum computer, a concept known as the no-cloning theorem [8]. These observations imply that any attempt by Eve to gain information about the key results in a perturbation that can be observed from Bob's measurement of the quantum states. Hence, QKD protocols are often referred to as unconditionally secure, since the security is based on the fundamental principles of quantum mechanics, rather than Eve's limited power or resources.

Despite the promise that this technology has offered over classical cryptosystems, QKD did not attract much attention until 1989, when Bennett and Brassard demonstrated the first prototype QKD experiment over 32.5 cm in free space [9]. This simple proof-of-principle demonstration attracted the interest of many researchers, and over the next three decades, the field has evolved from simple laboratory-based demonstrations to commercial products. Prototype QKD systems have been implemented using various photonic degrees of freedom (such as polarization, time, phase, and frequency), using many different protocols, and over a wide range of length scales. Recent studies have also demonstrated QKD links at ultra-long distances, such as in Earth-to-Satellite (1200 km) free space links [10], as well as in optical fiber, extending up to 404 km [11]. These projects ultimately aim to connect any two cities around the world using a Satellite as a trusted node for communication, and to build large-scale networks of QKD links connecting metropolitan cities via existing digital communication infra-structure.

Although QKD is a proven technique that has immense potential for securing communication in the post quantum computing era, there are still many implementation challenges that need to be resolved to make this technology suitable for mainstream communication world-wide. QKD requires transmission of signal states, one photon at a time, through lossy fiber optic- or free space-based quantum channels. In free-space channels, the atmospheric transmission varies over 10–100 ms timescales, and is wavelength-dependent, which means the operating wavelength range for these systems is limited [12]. In addition, the collection efficiency of single photons in a free-space link is very low due to the background radiance or stray light photons [12]. Similarly, in a fiber implementation, the quantum channel has a coefficient of loss of 0.2 dB/km (at 1550 nm wavelength), which means many transmitted photons are lost in the channel due to absorption or scattering. Unlike classical optical communication, the signal strength in QKD cannot be amplified using an optical amplifier without disturbing the quantum states, a direct implication of the no-cloning theorem [8]. Although there is a significant effort from the community to demonstrate repeaters that can boost quantum signals, no such technology currently exists. The absence of quantum repeaters means that there is a fundamental trade-off between the rate and distance for any QKD system operating in fiber optic links.

Another major challenge for all QKD systems is that the maximum rate at which single-photons can be detected in a QKD receiver is significantly lower than the rate at which quantum states can be prepared. The limitation in the detection rate is mainly due to the long recovery time of the single photon detectors, also known as the deadtime, over which a single-photon detector remains unresponsive to the incoming photons. Most of the current single-photon counting modules have a deadtime that is greater than 10–100 ns ($<$100 MHz), which means when two or more photons arrive at the detector in a time interval shorter than the deadtime, the detector cannot detect the second photon. In comparison, classical communication systems that use high-speed photoreceivers can detect photons at rates exceeding 10s of gigahertz. Therefore, the rate at which a secret key is generated in QKD is orders-of-magnitude lower than that of rates generated in digital communication [13].

Furthermore, most QKD systems are not compatible with current digital communication infrastructure, mainly due to the various degrees of freedom that are used to encode information in different QKD protocols. While some protocols use time and phase degrees of freedom to encode information, there are other protocols that use polarization and orbital-angular momentum degrees of freedom. Integration of QKD systems into existing platforms will require a major overhaul of the current transmitters and receivers, including replacing the classical photoreceivers with single-photon detectors, some of which will require cryogenic-based operations.

My contribution to the field of QKD is primarily in the development of novel high-speed QKD protocols that encode multiple bits of information per state, using high-dimensional time and phase degrees of freedom of single photon states. The aim of this thesis is to discuss the physics behind these novel QKD protocols that promise to mitigate most of these implementation challenges. In addition, I analyze the security of all the proposed protocols using analytical and numerical techniques. This thesis is a discussion of these projects, organized from the fundamental ideas of qubit-based ($d = 2$) QKD protocols (Chap. 2) to more advanced high-dimensional ($d > 2$) protocols in the later chapters.

1.1 Novel Contribution and Outline

In this section, I consider a simple example, based on the specifications of a commercial QKD system [14], to illustrate the need for novel QKD protocols that can generate high secret key rates. I show that the rate at which this system generates a secret key is not feasible for mainstream secure communication, and that the rate is ultimately limited by the long detector recovery time. I then outline the remaining chapters of this thesis with a brief summary of all the accomplishments.

Consider a QKD system that generates quantum states at a repetition rate of 500 MHz,[1] and transmits the quantum states through a quantum channel of transmission 0.01, which is equal to 100 km distance (20 dB loss) in a standard telecommunication fiber of coefficient of loss 0.2 dB/km. Assuming a mean photon number of 0.1 photon/state, and a detector efficiency of 50%, the rate of photon arrival is given by $R \approx 0.01 \times 0.1 \times 0.5 \times 500 = 0.25$ MHz. Given that each photon encodes a maximum of one bit of information, this rate (0.25 Mbps) is about a factor of 50 smaller than the average speed of an internet service provider in the USA [15]. In the opposite limit, for example, 20 km distance in fiber corresponding to a transmission of 0.4, the rate of photon arrival is $R \approx 0.4 \times 0.1 \times 0.5 \times 500 = 10$ MHz, which will start to saturate many single-photon counting modules that are available today.

To make QKD more relevant for widespread deployment in communication networks, there has been a significant push to increase the key generation rate of these systems, prioritizing metropolitan distances (20–80 km) for large-scale implementation of quantum networks [16]. One of the major breakthroughs was the development of superconducting nano-wire single-photon detectors (SNSPDs) that can detect single photons with high-efficiency (>90%) and yet have low dark count rates (10–100 counts/s) [17]. However, these detectors still have a recovery time greater than 10 ns [18], mainly limited by the readout electronics of the detectors.

One solution to the detector deadtime problem is to encode information in the high-dimensional state space of a photon. High-dimensional quantum states—qudits rather than qubits—provide a robust and efficient platform to overcome some of the practical challenges of current QKD systems [19, 20]. Fundamentally, QKD systems using a high-dimensional quantum state space have two major advantages over the qubit-based protocols. First, they can encode many bits ($\log_2 d$) of information on a single photon, and therefore increase the effective key generation rate in systems limited by the saturation of the single photon detectors. This becomes particularly important in the low loss quantum channels. Second, high-dimensional QKD systems have higher resistance to quantum channel noise, which means these systems can tolerate a higher quantum bit error rate, compared to qubit-based systems [21].

High-dimensional QKD protocols can be implemented using the same degrees of freedom as qubit-based protocols. For example, in the recent years, two popular choices have been the spatial modes, such as orbital angular momentum [22, 23] and the time-frequency (or phase) [24–29]. In Chap. 3, I discuss the implementation of a $d = 4$ time-phase QKD system that encodes information in the time basis, and detects the presence of an eavesdropper by transmitting states in the phase basis. In the past, such protocols were not implemented due to the complex transmitter

---

[1]These are the specifications of a standard QKD transmitter and receiver when this project was started in 2013–2014 [14].

and receiver that were needed to measure the phase basis states. The phase states can be prepared and measured using a combination of time-delays, optical switches and phase shifters, or using a tree of time-delay interferometers [24] as described in Chap. 3. Each photon received in this protocol can encode a maximum of 2 bits of information, which is a factor of two better than the qubit-based version. To enhance the secret key rate further, this system incorporates a receiver with four detectors in parallel to measure the quantum states in each basis, and thereby increases the raw detection rates by approximately a factor of four at short distances. This improved encoding and detection scheme allows the system to generate a secret key at megabits-per-second rates, from 4 dB channel loss corresponding to a distance of 20 km in standard fiber, up to a loss of 16.6 dB ($\sim$83 km in standard fiber). These rates are at least a factor of 2–5 times better than the current state-of-the-art QKD systems in the field [30].

In Chap. 3, I also analyze the security of this protocol considering many experimental non-idealities, such as state preparation flaw, losses in measurement devices, quantum efficiency and dark counts of single-photon counting modules. Previously, many security proofs of QKD protocols were implemented assuming ideal state preparation and detection devices. However, in the last decade, numerous studies have shown that such ideal assumptions open new attack strategies for eavesdroppers, also known as security loopholes. Such attack strategies were demonstrated to be detrimental to the security of many commercial QKD schemes [31, 32]. In this work, I not only take these experimental non-idealities into account, but also analyze the security of this protocol in the limit where a finite length of key is transmitted between the sender and receiver. As was demonstrated in many qubit-based protocols [33], when the length of the key is less than 10^6-bit long, the so-called finite-key effect reduces the secret key rate of these systems significantly. Prior to this work, no other discrete-variable high-dimensional QKD protocol has considered the finite-key contributions.

While the $d = 4$ QKD system described in Chap. 3 can generate a secret key at a high-rate, the protocol requires generation of four states in the time basis and four states in the phase basis. In Chap. 4, I discuss an efficient and simple high-dimensional platform, which makes the time-phase QKD system of Chap. 3 more practical for field implementation. Specifically, using numerical optimization, I show that some of the states generated in the system to monitor the presence of an eavesdropper are redundant, and therefore, are not required for the security of the protocol. Previously, such a study was performed only for $d = 2$ QKD systems using an analytic technique [34]. However, no studies explored the possibility of extending the technique of Ref. [34] to $d > 2$ QKD protocols. In addition, when the analytical technique proposed in Ref. [34] is applied to higher dimensional protocols, only one of the d states is found to be redundant, that is, one still needs to send $d - 1$ monitoring states to prove the security of an arbitrary d-dimensional protocol. In this study, I generalize the result and show that the redundancy of monitoring states goes beyond just one state. In fact, a high-dimensional QKD system can be implemented using just one monitoring basis state. These findings have significant implications in the design and implementation of QKD transmitters. For example, using this proof and a simplified experimental setup, I show that the secret key rate of the simplified $d = 4$ time-phase protocol is comparable to the full setup. The small reduction in the secret key rate simplifies the experimental setup significantly, making the $d = 4$ time-phase setup very practical for field implementation.

In Chap. 5, I use the result of Chap. 4 and demonstrate a new technique for measuring high-dimensional phase states using a local ancilla qudit (quantum-controlled) at Bob's receiver. This new receiver design does not require any time-delay interferometers like in the previous setups. Previously, the measurement of a complete set of d-dimensional phase states required $d - 1$ time-delay interferometers. The implication of this work is that it simplifies the phase state measurement device of the high-dimensional QKD system described in Chaps. 3 and 4 to just one beamsplitter and a second source of ancilla qudits, for any arbitrary dimension. I demonstrate an experimental proof-of-principle demonstration of this protocol, and analyze the security of this scheme.

In Chap. 6, I investigate the possibility of an eavesdropper attacking the high-dimensional phase states described in Chaps. 3–5 using a universal optimal cloning device. The device is constructed using linear optics and a laser source that can be used to clone the quantum states between Alice and Bob. As mentioned previously, the no-cloning theorem forbids cloning of unknown quantum states. However, there is an optimal limit up to which cloning of quantum states is possible [35–37]. To do so, one requires a source of quantum states that generates exactly one photon per state. Practical sources of quantum states, such as a phase randomized weak coherent source generates photon number per state determined by a Poisson distribution. Therefore, some of the states have zero or more than one photon per state. To extract the cloning fidelity of the single-photon state, I use the decoy-state technique developed in Chaps. 3–5, and place tight bounds on the cloning fidelity for high dimensional states. The implications of this work are twofolds. First, it shows that one can extract single-photon behavior using the decoy-state method developed in this thesis. Second, it shows that the cloning fidelity of high-dimensional quantum states decreases as a function of dimension, thereby providing a direct evidence of the robustness of high-dimensional quantum states.

A summary of all of my projects, the new physics and contribution that each project had to the field of QKD is given in Fig. 1.1.

Novel Idea in this thesis	Impact/New Physics	Comparison with Previous Works	Publications
High-dimensional time-phase QKD (Ch. 3)	Record-setting secret key rates at quantum channel losses equivalent to 20-83 km distances in standard optical fiber; first demonstration of finite-key security for discrete-variable high-dimensional QKD protocols	Higher secret key rates than any previous QKD systems with finite-key security proof	N. T. Islam *et al.* Sci. Adv. **3**, e1701491 (2017). N. T. Islam *et al.* Phys. Rev. Appl. **7**, 044010 (2017).
Efficient encoding scheme for a *d*-dimensional QKD system (Ch. 4)	Significantly simplifies the transmitter designs of arbitrary dimensional QKD systems; first demonstration of generalized unstructured QKD schemes for high-dimensional protocols	Previously only done for $d = 2$ protocols	N. T. Islam *et al.* Phys. Rev. A **97**, 042347 (2018).
High-dimensional time-phase QKD with quantum controlled measurement (Ch. 5)	Greatly simplifies the phase state measurement device of high-dimensional time-phase QKD systems; first demonstration of quantum-controlled measurement scheme for phase state measurement	No previous demonstration	
Optimal quantum cloning of high-dimensional phase states (Ch. 6)	A novel technique to estimate the performance of pure single-photon states using weak coherent states; first application of decoy-state technique in a quantum cloning experiment	Previously done for only orbital angular momentum states and with single-photon states	N. T. Islam *et al.*, in preparation (2018).

Fig. 1.1 List of novel findings. This table lists a brief summary, the new physics and impact that each of my contributions has made in the field of QKD

References

1. S. Debnath, N. Linke, C. Figgatt, K. Landsman, K. Wright, C. Monroe, Nature **536**, 63 (2016)
2. N.M. Linke, D. Maslov, M. Roetteler, S. Debnath, C. Figgatt, K.A. Landsman, K. Wright, C. Monroe, Proc. Natl. Acad. Sci. U. S. A. **114**, 3305 (2017). http://dx.doi.org/10.1073/pnas.1618020114. http://www.pnas.org/content/114/13/3305.full.pdf
3. T.S. Metodi, D.D. Thaker, A.W. Cross, in *Proceedings of the 38th Annual IEEE/ACM International Symposium on Microarchitecture, MICRO 38* (IEEE Computer Society, Washington, 2005), pp. 305–318. http://dx.doi.org/10.1109/MICRO.2005.9

4. M. Ahsan, R.V. Meter, J. Kim, J. Emerg. Technol. Comput. Syst. **12**, 39:1 (2015). http://dx.doi. org/10.1145/2830570

5. P.W. Shor, in *Proceedings of 35th Annual Symposium on Foundations of Computer Science* (1994), pp. 124–134. http://dx.doi.org/10.1109/SFCS.1994.365700

6. S. Barnett, *Quantum Information*. Oxford Master Series in Physics (OUP, Oxford, 2009). https://books.google.com/books?id=A2k4HH2tFR8C

7. C.H. Bennett, G. Brassard, in *1984 International Conference on Computers, Systems & Signal Processing*, Bangalore (1984), pp. 175–179

8. W.K. Wootters, W.H. Zurek, Nature **299**, 802 (1982)

9. C.H. Bennett, G. Brassard, ACM Sigact News **20**, 78 (1989).

10. S.-K. Liao, W.-Q. Cai, W.-Y. Liu, L. Zhang, Y. Li, J.-G. Ren, J. Yin, Q. Shen, Y. Cao, Z.-P. Li, F.-Z. Li, X.-W. Chen, L.-H. Sun, J.-J. Jia, J.-C. Wu, X.-J. Jiang, J.-F. Wang, Y.-M. Huang, Q. Wang, Y.-L. Zhou, L. Deng, T. Xi, L. Ma, T. Hu, Q. Zhang, Y.-A. Chen, N.-L. Liu, X.-B. Wang, Z.-C. Zhu, C.-Y. Lu, R. Shu, C.-Z. Peng, J.-Y. Wang, J.-W. Pan, Nature **549**, 43 EP (2017)

11. H.-L. Yin, T.-Y. Chen, Z.-W. Yu, H. Liu, L.-X. You, Y.-H. Zhou, S.-J. Chen, Y. Mao, M.-Q. Huang, W.-J. Zhang, H. Chen, M.J. Li, D. Nolan, F. Zhou, X. Jiang, Z. Wang, Q. Zhang, X.-B. Wang, J.-W. Pan, Phys. Rev. Lett. **117**, 190501 (2016). http://dx.doi.org/10.1103/PhysRevLett. 117.190501

12. R.J. Hughes, J.E. Nordholt, D. Derkacs, C.G. Peterson, New J. Phys. **4**, 43 (2002). http://stacks. iop.org/1367-2630/4/i=1/a=343

13. H.-K. Lo, M. Curty, K. Tamaki, Nat. Photonics **8**, 595 (2014)

14. ID Quantique. https://www.idquantique.com/

15. Akamai, *State of the Internet, Q1* (Akamai Technologies, Cambridge, 2017)

16. E. Diamanti, H.-K. Lo, B. Qi, Z. Yuan, npj Quantum Inf. **2**, 16025 EP (2016)

17. F. Marsili, V.B. Verma, J.A. Stern, S. Harrington, A.E. Lita, T. Gerrits, I. Vayshenker, B. Baek, M.D. Shaw, R.P. Mirin et al., Nat. Photonics **7**, 210 (2013)

18. Q. Zhao, T. Jia, M. Gu, C. Wan, L. Zhang, W. Xu, L. Kang, J. Chen, P. Wu, Opt. Lett. **39**, 1869 (2014). http://dx.doi.org/10.1364/OL.39.001869

19. H. Bechmann-Pasquinucci, W. Tittel, Phys. Rev. A **61**, 062308 (2000). http://dx.doi.org/10. 1103/PhysRevA.61.062308

20. N.J. Cerf, M. Bourennane, A. Karlsson, N. Gisin, Phys. Rev. Lett. **88**, 127902 (2002). http:// dx.doi.org/10.1103/PhysRevLett.88.127902

21. L. Sheridan, V. Scarani, Phys. Rev. A **82**, 030301 (2010). http://dx.doi.org/10.1103/PhysRevA. 82.030301

22. J. Leach, E. Bolduc, D.J. Gauthier, R.W. Boyd, Phys. Rev. A **85**, 060304 (2012). http://dx.doi. org/10.1103/PhysRevA.85.060304

23. M. Mirhosseini, O.S. Magaa-Loaiza, M.N. OSullivan, B. Rodenburg, M. Malik, M.P.J. Lavery, M.J. Padgett, D.J. Gauthier, R.W. Boyd, New J. Phys. **17**, 033033 (2015). http://stacks.iop.org/ 1367-2630/17/i=3/a=033033

24. T. Brougham, S.M. Barnett, K.T. McCusker, P.G. Kwiat, D.J. Gauthier, J. Phys. B **46**, 104010 (2013). http://stacks.iop.org/0953-4075/46/i=10/a=104010

25. J. Nunn, L.J. Wright, C. Söller, L. Zhang, I.A. Walmsley, B.J. Smith, Opt. Express **21**, 15959 (2013). http://dx.doi.org/10.1364/OE.21.015959

26. J. Mower, Z. Zhang, P. Desjardins, C. Lee, J.H. Shapiro, D. Englund, Phys. Rev. A **87**, 062322 (2013). http://dx.doi.org/10.1103/PhysRevA.87.062322

27. D.J. Gauthier, C.F. Wildfeuer, H. Guilbert, M. Stipcevic, B.G. Christensen, D. Kumor, P. Kwiat, K.T. McCusker, T. Brougham, S. Barnett, in *The Rochester Conferences on Coherence and Quantum Optics and the Quantum Information and Measurement meeting* (Optical Society of America, New York, 2013), p. W2A.2. http://dx.doi.org/10.1364/QIM.2013.W2A.2

28. Z. Zhang, J. Mower, D. Englund, F.N.C. Wong, J.H. Shapiro, Phys. Rev. Lett. **112**, 120506 (2014). http://dx.doi.org/10.1103/PhysRevLett.112.120506

29. T. Brougham, C.F. Wildfeuer, S.M. Barnett, D.J. Gauthier, Eur. Phys. J. D **70**, 214 (2016). http://dx.doi.org/10.1140/epjd/e2016-70357-4

30. C. Lee, D. Bunandar, Z. Zhang, G.R. Steinbrecher, P.B. Dixon, F.N.C. Wong, J.H. Shapiro, S.A. Hamilton, D. Englund, High-rate field demonstration of large-alphabet quantum key distribution (2016). http://arxiv.org/abs/arXiv:1611.01139. arXiv:1611.01139
31. N. Jain, E. Anisimova, I. Khan, V. Makarov, C. Marquardt, G. Leuchs, New J. Phys. **16**, 123030 (2014). http://stacks.iop.org/1367-2630/16/i=12/a=123030
32. L. Lydersen, C. Wiechers, C. Wittmann, D. Elser, J. Skaar, V. Makarov, Nat. Photonics **4**, 686 (2010).
33. M. Tomamichel, C.C.W. Lim, N. Gisin, R. Renner, Nat. Commun. **3**, 634 (2012)
34. K. Tamaki, M. Curty, G. Kato, H.-K. Lo, K. Azuma, Phys. Rev. A **90**, 052314 (2014). http://dx.doi.org/10.1103/PhysRevA.90.052314
35. V. Bužek, M. Hillery, Phys. Rev. A **54**, 1844 (1996). http://dx.doi.org/10.1103/PhysRevA.54.1844
36. R.F. Werner, Phys. Rev. A **58**, 1827 (1998). http://dx.doi.org/10.1103/PhysRevA.58.1827
37. N. Gisin, S. Popescu, Phys. Rev. Lett. **83**, 432 (1999). http://dx.doi.org/10.1103/PhysRevLett.83.432

Chapter 2
Building Blocks of Quantum Key Distribution

The first QKD scheme proposed by Bennett and Brassard in 1984 was a qubit-based ($d = 2$) protocol that encodes information using a photon's polarization modes in two non-orthogonal bases [1, 2]. Polarization is one of the many different degrees of freedom that can be used to encode information in a QKD system. The information can also be encoded using time-phase [3–5], orbital angular momentum [6], etc. degrees of freedom. In this chapter, as an introductory discussion and motivation for high dimensional QKD, I describe an ideal $d = 2$ time-phase QKD system.

Soon after the discovery and demonstration of the first QKD protocol, researchers in the field realized that the idea of unconditional security cannot apply to a practical QKD system. An underlying assumption in the original proposal and proof of unconditional security is that QKD systems can generate and measure ideal quantum states, which is difficult to accomplish even with the state-of-the-art equipment available today. An eavesdropper (Eve) can exploit the imperfect experimental realizations and implement eavesdropping strategies based on the discrepancy between the theoretical models, and the actual implementation of the systems. Instead, researchers in the field are now focused on provable security of QKD protocols that take into account the non-ideal behavior of experimental setups. A large portion of this thesis is devoted to minimizing this gap between theory and experimental non-idealities, often through stringent conditions in the development of security proofs that give a conservative estimate of the expected secret key rate, while ensuring highest level of security.

Nevertheless, in this chapter, I analyze the security of an ideal $d = 2$ time-phase system, and provide an intuitive explanation why such a protocol is secure against a simple intercept-and-resend eavesdropping strategy by Eve. I also discuss the building blocks required to realize a QKD system and the limitations of the $d = 2$ protocols. Furthermore, I discuss various techniques that can be used to increase the secret key rate of such a system, such as wavelength-division multiplexing and high-dimensional encoding.

© Springer Nature Switzerland AG 2018
N. T. Islam, *High-Rate, High-Dimensional Quantum Key Distribution Systems*,
Springer Theses, https://doi.org/10.1007/978-3-319-98929-7_2

2.1 Time-Phase QKD in $d = 2$ Hilbert Space

2.1.1 Time-Phase States and Mutually Unbiased Bases

The goal of any QKD system is to share a string of classical bit values, known as the secret key, between a sender (Alice) and a receiver (Bob) by transmitting quantum photonic states. In a typical qubit-based protocol, the quantum states are analogous to a two-level quantum system, where each level represents a different binary bit value and is orthogonal with respect to the other. Alice can prepare quantum states in one of these levels to encode the bit values '0' or '1'. Figure 2.1a illustrates an example, wherein the states $|0\rangle$ and $|1\rangle$ represent the bit values of '0' and '1', respectively, in the vertical/horizontal basis, also known as the computational basis.

A fundamental requirement for the security of QKD is that Alice also encodes information in a basis that is non-orthogonal and mutually unbiased with respect to the computational basis. One such basis is known as the diagonal/anti-diagonal basis (Hadamard basis) that encodes the bit values, '0' and '1', using the states $|+\rangle = (|0\rangle + |1\rangle)/\sqrt{2}$ and $|-\rangle = (|0\rangle - |1\rangle)/\sqrt{2}$ as illustrated in Fig. 2.1b. The idea of mutually unbiased basis (MUB) is that when a state is prepared in a given basis and measured in a MUB, the outcome of the measurement is a uniformly random bit value. For instance, consider measuring a Hadamard basis state in the computational basis. When any of the Hadamard basis states are measured in the computational basis, there is a 1/2 probability of obtaining either a $|0\rangle$ or a $|1\rangle$ state. Mathematically, this means that the squared-overlaps of the quantum states, $|\langle \pm |0\rangle|^2$, and $|\langle \pm |1\rangle|^2$ are 1/2. In other words, when states prepared in a given basis are measured in a MUB, an incorrect bit value occurs 50% of the time. This phenomenon is fundamental for the security of QKD systems.

In a time-phase QKD scheme, information is encoded in either the time (computational) or phase (Hadamard) basis. The classical analog of such a system is the pulse position modulation scheme, in which a classical photonic wavepacket (pulse) is localized within a time bin of width τ. A set of two contiguous time bins is known as a frame. Depending on where the pulse is positioned in a given frame relative to the first time bin, the pulse can encode either a bit value of '0' or a '1',

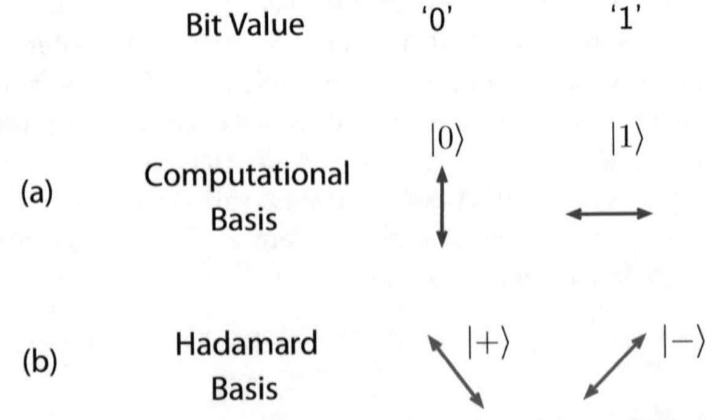

Fig. 2.1 Mutually unbiased bases. Two maximally non-orthogonal bases: (a) Computational basis, and (b) Hadamard basis

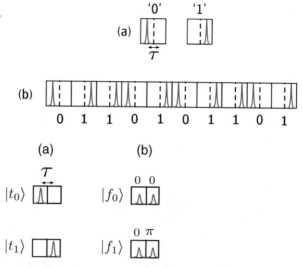

Fig. 2.2 Pulse position modulation. (**a**) Illustration of binary bit values '0' and '1' in a pulse position modulation scheme. (**b**) An example of a string of bit values encoded using a pulse position modulation scheme

0 1 1 0 1 0 1 1 0 1

Fig. 2.3 Time-phase states in $d = 2$. (**a**) The relative position of the wavepacket determines the state $|t_0\rangle$ and $|t_1\rangle$. (**b**) The phase basis states are superposition of the time-basis states with 0 and π phase difference between the first and the second peak of the single-photon wavepacket

as shown in Fig. 2.2a. A string of bit values can be encoded by placing pulses in a continuous string of frames as shown in Fig. 2.2b.

In case of time-bin QKD, the pulses are replaced with single photon wavepackets and the mean photon number is adjusted so that there is approximately one photon per frame. In $d = 2$, there are two such states, one in which the photon is placed in the first time bin, represented as $|t_0\rangle$, and the other in which the photon is placed in the second time bin, represented as $|t_1\rangle$. The basis set consisting of states $|t_0\rangle$ and $|t_1\rangle$ are time-bin analog of the computational basis, and are therefore referred to as the time-basis states. The same information can also be encoded in a conjugate basis, which is mutually unbiased with respect to the time bin states, such as

$$|f_0\rangle = \frac{1}{\sqrt{2}}(|t_0\rangle + |t_1\rangle),$$

$$|f_1\rangle = \frac{1}{\sqrt{2}}(|t_0\rangle - |t_1\rangle). \tag{2.1}$$

I refer to the basis consisting of states $|f_0\rangle$ and $|f_1\rangle$ as the phase basis, since the states are a quantum superposition of the time-basis states with a relative phase difference between the first and second peak as illustrated in Fig. 2.3b. It is the relative phase between the two peaks that encode information in phase basis.

The phase basis is not the only basis that is mutually unbiased with respect to the time basis. Another set of basis states are: $(|t_0\rangle + i|t_1\rangle)/\sqrt{2}$ and $(|t_0\rangle - i|t_1\rangle)/\sqrt{2}$. This set of basis states is useful in many implementations of QKD, such as the six-state protocol [7]. However, for the discussion below, I will consider only the time and the phase bases.

2.1.2 The Protocol

Suppose Alice wants to share an n-bit long classical random key with Bob. This is accomplished using a time-phase QKD protocol as follows.

Alice generates a random bit value using a quantum random number generator (QRNG). This is the first bit of the secret key that she wants to share with Bob. The use of QRNG is critical because the randomness of a QRNG is extracted from quantum processes that are believed to be truly random [8]. She also uses another QRNG to choose the basis (time or phase), and prepares the quantum state in that basis. She then transmits the quantum state to Bob via a quantum channel (free-space or optical fiber).

When the photon arrives at Bob's receiver, he uses another QRNG to determine the basis in which to measure the photon. Given that the QRNG picks time and phase bases with equal probability, there is a 50–50% chance that Alice's basis choice matches with Bob's, and therefore a correct state (bit value) is determined 50% of the time. Bob may also determine the basis using a beamsplitter, which passively directs the photon to either time or phase basis measurement device, allowing the quantum nature of photons to determine the basis choice.

After N rounds of transmission and reception, Alice and Bob communicate over an authenticated public channel, and Bob announces his basis choice for each round of transmission when he received a photon. A channel is referred to as authenticated when an eavesdropper can listen to the conversation but does not participate or provide "denial-of-service," such as disrupting the post-transmission conversation between Alice and Bob. An authenticated channel is usually calibrated at the beginning of the protocol by Alice and Bob.

After Alice and Bob discuss their basis choices, they only keep the events from the rounds where their basis choices are identical, and discard the remaining. In principle, they now possess a perfectly correlated string of key, commonly referred to as the sifted key. However, in practical implementations this is often not true as the imperfections in state preparation, detection, and noise in the quantum channel introduce random errors to the quantum signals, resulting in quantum bit errors. Such errors might also result from Eve trying to attack the quantum channels to gain information.

To ensure that the final secret key string is independent of error bits, three additional steps are performed, often referred to as post-processing steps. First, a random fraction of the secret key is announced, which is used to estimate the error rates in the sifted key. This is known as parameter estimation. This fraction should be large enough so that a good estimate of the error rate is obtained with low statistical uncertainty. Second, a classical error correction algorithm is implemented on the remaining fraction. Error correction is a way to correct any errors in the remaining key bit values by broadcasting a small fraction of the key. The size of the key that is announced during error correction depends on the error rate determined from the parameter estimation step. Finally, a privacy amplification algorithm is implemented to eliminate any correlation that the key may have with an eavesdropper, thereby making the final key completely secret.

2.1.3   Entanglement-Based vs. Prepare-and-Measure Protocols

The QKD protocol discussed above is a form of prepare-and-measure scheme where Alice prepares a quantum state and Bob performs a random measurement on the quantum state. An equivalent yet different approach to QKD protocols involves entanglement, in which Alice prepares two entangled photons of the form [7]

$$|\phi^+\rangle = \frac{1}{\sqrt{2}} \left(|t_0\rangle_A |t_0\rangle_B + |t_1\rangle_A |t_1\rangle_B \right), \tag{2.2}$$

where the subscripts A and B on the quantum states represent Alice and Bob, respectively. In this scheme, Alice keeps one of the entangled photons, and transmits the other one to Bob. In the ideal scenario, they are in possession of a perfectly correlated photon pair. This means, when Alice performs a measurement on her photon, she can immediately determine the state that Bob has received. To encode information in the phase basis, Alice can perform a unitary operation on the entangled photons to prepare photons in the mutually unbiased phase basis

$$|\phi^+\rangle = \frac{1}{\sqrt{2}} \left(|f_0\rangle_A |f_0\rangle_B + |f_1\rangle_A |f_1\rangle_B \right). \tag{2.3}$$

Broadly speaking, both the prepare-and-measure and entanglement-based approaches are equivalent in that one can be translated into other. Alice's measurement on the quantum states determine the state that Bob will receive, and therefore this becomes equivalent to the prepare-and-measure scheme.

From Bob's perspective, the two approaches are identical since the information is still encoded on single-photon states. However, from the practical point of view, there is a trade-off between the two approaches. For example, sources of entangled photons such as heralded spontaneous parametric down conversion generate entangled states with high heralding efficiencies, and are therefore very reliable source of entangled photons [9]. However, the rate at which photon pairs can be generated is low compared to what can be achieved in a prepare-and-measure scheme with off-the-shelf equipment. Since the main goal of this thesis is to demonstrate high-rate QKD protocols, all work presented here will be using a prepare-and-measure scheme.

2.2 An Intuitive Approach to Security

Security analyses of QKD protocols require understanding of theoretical tools, such as the entropic uncertainty principle [10] or entanglement distillation [7]. Such proofs are presented in Chaps. 3–5. Here, I consider a simple attack strategy by Eve, and analyze how this attack can introduce error in the final key shared between Alice and Bob, thereby revealing the presence of Eve in the quantum channel.

Consider a 20-bit long key string that Alice wants to share with Bob in the presence of an eavesdropper, as illustrated in Fig. 2.4a. Suppose Alice's basis choice for each round of transmission is random, as shown in Fig. 2.4b. After the transmission of a quantum state, Eve intercepts and measures the state in a random basis to determine the bit value encoded on the photon. This type of attack is known as an intercept-and-resend attack.

If Eve decides to intercept and measure every photon in the channel, her random basis choice will match with Alice's basis 50% of the time by chance. In other words, out of 20 bits, she will have complete knowledge of 10 bits. For the remaining cases where she measures the states in the wrong basis, she will still guess the correct bit value 50% of the time, since states prepared in a given basis and measured in a MUB result in a correct (or incorrect) outcome 50% of the time. Therefore, among the remaining 10 bits when she guesses the incorrect basis, she obtains correct bit values for 5 bits. This is illustrated in Fig. 2.4b where the incorrect basis and bit values are marked red.

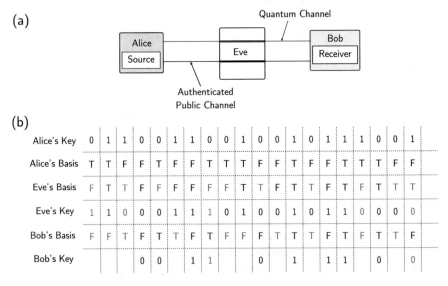

Fig. 2.4 Intercept-and-resend eavesdropping strategy. An illustration of how the intercept-and-resend attack strategy by an eavesdropper results in 25% quantum bit error rate in the sifted key

After Eve intercepts and measures the states in the quantum channel, she generates and transmits new qubits to Bob using the same basis choices that she used to measure the photons. When the photons arrive at Bob's receiver, he chooses to measure the quantum states in either time or phase basis without knowing that Eve has interacted with the photons in the quantum channel. Since the basis choices for Alice and Eve are random, there is a 50% chance that Bob will measure the states in the same basis in which Alice/Eve encoded the information. However, the bit values that Bob measures for those events will not always correlate with Alice, since Eve's interaction with the quantum states compels Eve to encode wrong bit values for 25% of the events. For the 10 events where the basis choices for Alice and Bob match, there are now 2.5 bits on average that are incorrect (Fig. 2.4b), which is revealed during the parameter estimation step of the protocol.

It is important to note that the errors have resulted precisely due to Alice encoding information in two mutually unbiased non-orthogonal bases, and Eve's random guessing of Alice's basis choice. In many implementations of QKD protocols, states in one basis are used primarily to determine the presence of an eavesdropper, thus are called the monitoring-basis states. The quantum bit error rate in that basis gives an estimate of how much information is leaked to an eavesdropper.

2.3 Eavesdropping Strategies

The intercept-and-resend attack discussed above is a technique that an eavesdropper may employ to gain information about the encoding qubits transmitted between Alice and Bob. However, this is a very specific and weak strategy since it introduces a large error rate (25%) in the final key shared between Alice and Bob. In general, QKD protocols assume that an eavesdropper have all the resources and power within the laws of Physics. Therefore, Eve might employ technologies that are not available today, but are still allowed within the laws of Physics, such as a quantum memory that can hold quantum states indefinitely.

With the power of Eve restrained only by the laws of Physics, all QKD security proofs must specify the level of eavesdropping attack against which the protocols are secure. A QKD protocol that is secure against an intercept-and-resend attack might be vulnerable to a stronger eavesdropping strategy. Therefore, it is important to classify the eavesdropping strategies and define the level of security for each protocol. Below I summarize the common types of eavesdropping strategies [7].

Independent (or Incoherent) Attacks

Independent attacks are the type of eavesdropping strategies whereby an eavesdropper attacks each qubit independently, and interacts with each qubit using the same strategy. Additionally, she performs the measurement of the quantum states prior to the classical post-processing step of the protocol. Intercept-and-resend is an example of an independent attack since Eve interacts with each qubit and measures the qubits prior to the post-processing step.

Another example of a stronger independent eavesdropping strategy is the photon number splitting (PNS) attack that an eavesdropper performs on a multi-photon quantum state. In a typical QKD implementation, quantum states are generated by modulating a coherent light source and then attenuating the states so that there is, on average, 1 photon per pulse. A coherent source emits photons based on a Poisson distribution, such that the probability of having n photons in a state generated with mean photon number μ is given by

$$P(n) = \frac{\exp(-\mu)\mu^{n}}{n!}. \tag{2.4}$$

Figure 2.5a shows the probability distributions as a function of the photon number for several mean photon numbers. The red, blue, and black lines correspond to $\mu = 0.1, \ 0.5, \ 1$, respectively. The plot illustrates that as μ increases the probability of

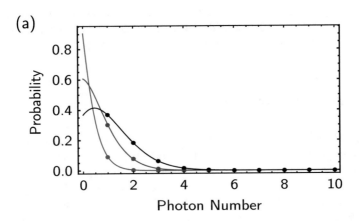

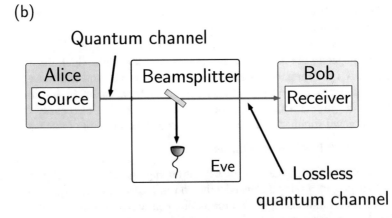

Fig. 2.5 Photon number distribution and photon number splitting attack. (**a**) Photon number distribution for the mean photon numbers $\mu = 0.1$ (red), 0.5 (blue), and 1 (black). (**b**) A schematic illustration of a photon number splitting attack

getting a state with more than one photon also increases. When $\mu = 0.1$, there is a 90.5% chance of getting a state with no photons (vacuum) and a 9.0% chance of getting a state with one photon. The probability of getting more than 1 photon is 0.5%.

An eavesdropper can potentially perform a PNS attack on the multi-photon states using an experimental setup such as the one shown in Fig. 2.5b. The PNS attack is one where an eavesdropper split one of the photons from a multi-photon state either using a beamsplitter or an optical switch. She can then measure the photon in one of the two bases, without introducing any error in the final key. Moreover, to avoid detection, Eve might replace the quantum channel with a lossless fiber so that the reduction in channel transmission due to her equipment is not detected. In this case, an eavesdropper may gain complete information about the multi-photon states without increasing the error rate in Bob's measurement device. Such an attack can be prevented using a technique known as decoy-state method, where Alice transmits the quantum states with different mean photon numbers to place tight bounds on the single-photon error rates. This is discussed in the next section and also in the next chapters in various contexts.

Collective Attacks

Collective attacks are similar to independent attacks in that an eavesdropper attacks each qubit and employs the same strategy. However, it is stronger than an independent attack because an eavesdropper can have a quantum memory and hold the qubits in the memory until the classical post-processing step between Alice and Bob.

A PNS attack is stronger when performed with a quantum memory. If Eve aggregates photons from the multi-photon states during the communication session, and performs the measurements only after Alice and Bob discuss their basis choices, then Eve can choose bases that match with those of Alice and Bob. In this case, Eve will know twice as many bits compared to the independent PNS attack, without introducing any error in the final key.

Coherent Attacks

Coherent attacks are the most general and flexible attack that an eavesdropper may perform on the quantum states. For example, she may adopt her attack strategies based on the prior measurements. She might entangle many quantum states and hold them in a quantum memory. Due to the flexibility of this type of attacks, it is often impossible to optimize the mutual information between Alice or Bob and Eve for such strategies. Therefore, a large number protocols in the literature are secure against an eavesdropper attacking the system using collective attack strategies. However, for most protocols, the security against collective attacks can be extended to more general coherent attacks using theoretical tools such as the quantum de Finnetii theorem [7].

Side-Channel Attacks or Quantum Hacking

Quantum hacking refers to eavesdropping attacks that exploit implementation flaws of a QKD system. An example of such an attack is the time-shift attack [11] in which an eavesdropper exploits the efficiency mismatch of Bob's single photon detectors to guess Bob's basis choice. Other examples of side-channel attacks include the Trojan horse attack [12] and the detector blinding attack [13]. In the Trojan horse attack, an eavesdropper shine a bright laser beam into Alice's encoding device and measures the reflected photons to gain information about the secret key. In contrast, the detector blinding attack exploits the properties of single-photon counting modules, such as the avalanche photodiodes (APDs), to constraint Bob to pick the same basis as the eavesdropper.

Recently, the detector blinding attacks have attracted a lot of interest since most commercial QKD systems use APDs, and are particularly vulnerable to such attacks. In a QKD system, APDs are run in gated mode so that the detector is "active" only when an event is expected. Detectors are gated to ensure that any unwanted events, such as dark counts, do not cause detection events when a qubit is expected from Alice. During the period when the detectors are in the "off" state, they are in the linear regime, which means the current output of the detectors is directly proportional to the input optical power. Therefore, the detectors behave like classical photodiodes.

To exploit this detector property, an eavesdropper can perform an intercept-and-resend attack. However, instead of sending quantum states, Eve sends classical light (just above the threshold power) so that if Bob chooses the same basis as Eve, the detector will 'click' like a real detection event. A 'click' is defined as a detector signal (pulse) that crosses a threshold voltage, so that it can be categorized as a registered event. When Eve does not choose the same basis as Bob, the classical pulse does not have enough power to make the detectors click [13]. This will result in half the events to be recorded as inconclusive events. In general, the transmission of quantum channel is much lower than 0.5, and therefore will not affect the system significantly.

There are many other types of side-channel attacks that have been demonstrated in practical QKD systems. The discussion of all these attacks is beyond the scope of this thesis. Typically, once a new side-channel attack is known, the remediation is straightforward, and usually involves hardware fixes, such as changing the QKD transmitter or receiver. However, this also means that the real threat on practical QKD system comes from the attacks that are not yet known.

Due to the threat from unknown side-channel attacks, in the recent years, researchers in the field have shown great interest in the development of new types of QKD protocols, such as the measurement device-independent (MDI) protocols [14–16]. The primary assumption in the MDI schemes is that the measurement device is untrusted, and Eve may potentially perform the measurements. However, in order to extract the information about Alice and Bob's quantum states, she has to perform a time-reversal Bell state measurement. Because she does not know the basis and the states of the incoming photons, she is unable to extract any additional information about the quantum states. Although some of the recent experiments have demonstrated the feasibility of these protocols [17–20], the secret key generation rates of these schemes are much lower than conventional QKD systems employing the prepare-and-measurement schemes. For the QKD protocols discussed in this thesis, I implement the systems using prepare-and-measure schemes, taking into account all the known security loopholes.

2.4 Practical Implementation of a $d = 2$ Time-Phase QKD System

In this section, I discuss the main components required to implement a time-phase QKD system. Specifically, I discuss how single-photon states in time and phase basis can be prepared at high-rate. Then, I discuss the measurement devices required to detect these states.

2.4.1 State Preparation

There are many ways to design and implement a QKD transmitter. For example, there are several options for the laser source, operating wavelength, and modulation scheme. Some sources such as a vertical-cavity surface-emitting laser (VCSEL) can be modulated directly with an rf-signal to generate the quantum states, while others, like a continuous-wave (cw) laser, require electro-optic intensity modulators. In addition to the simplicity of the design, a QKD transmitter must be developed considering several key criteria.

First, to avoid side-channel attacks, a practical QKD transmitter must generate quantum states that are very close to the ideal single-photon states. In other words, the quantum states in the time and phase bases must be perfectly mutually unbiased with respect to each other. In many early implementations of QKD protocols, non-ideal quantum state preparations led to security loopholes [7]. Second, the transmitter should be dynamic in that several QKD protocols can be implemented with one transmitter without overhauling the entire setup. Third, the quantum states must be prepared and transmitted at high-rate, preferably close to the classical communication standard. Fourth, a QKD transmitter must be easily integrable into existing classical communication infrastructure.

Based on all these criteria, a QKD transmitter can be designed using high-bandwidth electro-optic modulators. Specifically, an intensity modulator (IM) can be used to modulate a cw laser into narrow-width discrete optical wavepackets, and a phase modulator (PM) can be used to impose the discrete phase values as shown in Fig. 2.6. An important specification for the intensity modulators used in a QKD transmitter is that they must have high extinction ratio, which refers to the

Fig. 2.6 A QKD
transmitter. A high-level
experimental setup that can
be used to generate the $d = 2$
time-phase states

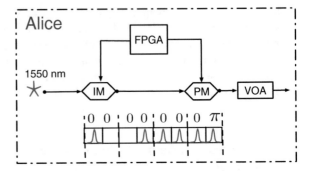

signal-to-background (noise) ratio of the optical wavepackets. In a QKD system, any uncorrelated photons result in a quantum bit error. Therefore, to avoid accidental detection events in wrong time bins, a high-suppression of the background light is required. These high-extinction ratio intensity modulators can be either custom-made by vendors or multiple intensity modulators can be placed in series to enhance the signal-to-background ratio.

The rf-signal used to drive the intensity modulator can be generated using a field programmable gate array (FPGA). To maximize the key generation rate, the intensity modulator can be driven at a clock rate >10 GHz. In principle, a random pattern that resembles Alice's basis and state choices can also be stored on the FPGA memory and then read out to generate the states. Finally, the optical wavepackets are attenuated using a variable optical attenuator (VOA) to ensure that each state contains one photon on average.

To overcome loss-dependent eavesdropping strategies such as the photon number splitting attack that arise from multi-photon components of the weak coherent pulses, the so-called decoy-state method can be implemented. As mentioned in the previous section, decoy-state technique requires Alice to send quantum states of varying intensity levels, randomly chosen so that an eavesdropper cannot guess the mean photon number per state. By observing the detection statistics for each intensity level, Alice and Bob can place tight bounds on the total number of single-photon events that are shared between them, as well as the error that an eavesdropper might have introduced by attacking the quantum states [21, 22].

To create the quantum states with different mean photon numbers, independent signals from the FPGA can be used to drive additional intensity modulators (not shown in Fig. 2.6). The amplitudes of the FPGA signals are adjusted to create quantum states with the required intensity levels (mean photon numbers). Frequently in QKD, vacuum states, corresponding to a mean photon number of zero, are used as one of the decoy intensities. These states can be generated by biasing an intensity modulator at the fringe maximum and adjusting the amplitude of the FPGA signal so that the transmission of light through the intensity modulator is momentarily blocked for a short duration when vacuum states are transmitted.

Finally, an important criterion for the security of any QKD system is that the global phase of the quantum states (frames) must be randomized. Global phase refers to the overall phase of a quantum state; it is different from the local phase used to define the phase states in Fig. 2.3b. Since the global phase value is arbitrary and same for all the time bins in a given frame, typically it is not explicitly mentioned while defining the quantum states. However, it is important to randomize this phase for each quantum state, so that there is no phase coherence between successive quantum frames. The phase randomization can be done in two different ways. First, the injection current of the cw laser can be modulated with an arbitrary shaped pulse, in synchronization with the FPGA signals that drive the intensity modulators. Second, an external phase modulator can be driven with an arbitrary step-like function to generate the discrete randomized phase values for each frame. A recent study of the second method showed that in a typical implementation, it is sufficient to randomize the phase values to ~ 10 discrete levels [23].

2.4.2   State Detection

At the receiver, the quantum states in time basis can be detected by using a single-photon counting module and a high-speed time-to-digital converter. Detection of the phase states can be performed by appropriately delaying the temporal wavepackets within a frame and interfering them in the same temporal mode. This can be accomplished by using active optical switches, optical delay lines, and phase shifters, although at the expense of a more complicated experimental setup [24]. An alternative method involves using a passive unequal path (delay) interferometers (DI) as shown in Fig. 2.7a.

In a time-delay interferometer (Fig. 2.7b), an incoming beam is split equally by a 50–50 beamsplitter and directed along two different paths. They are recombined at a second 50–50 beamsplitter where the wavepackets interfere. The difference in path between the two arms of the interferometer is denoted by $\Delta L = \Delta L_0 + \delta L$,

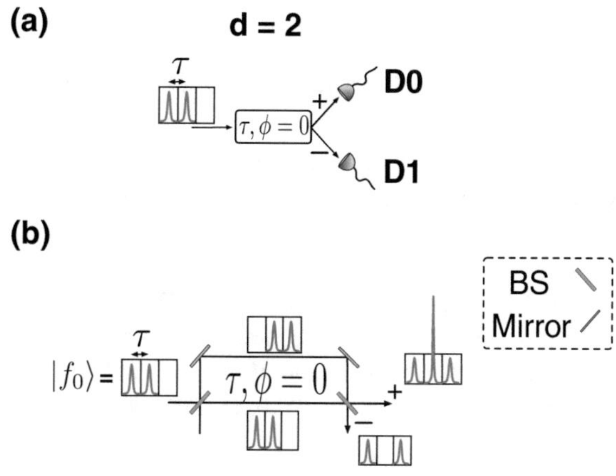

Fig. 2.7 Measurement scheme for $d = 2$ phase states. (**a**) A time-delay interferometer with a time-delay matched to the time-bin width τ and the phase set to 0. (**b**) A detailed illustration showing the propagation of the wavepacket as it travels through the long and short arm of the interferometer

where ΔL_0 is the nominal path difference. Here, $\delta L \ll \Delta L_0$ is a small path difference that allows to make a fine adjustment to the transmission resonances of the interferometer and is proportional to the phase $\phi = k\delta L$, where k is the magnitude of the wavevector of the wavepacket.

For $d = 2$, only a single time-delay interferometer is required with $\Delta L = c\tau$, corresponding to a free-spectral range (FSR) $c/\Delta L$, where c is the speed of light and ϕ is set to zero. When the state $|f_0\rangle$ is incident on the interferometer, the wavepacket traveling along the long path is delayed by τ with respect to the wavepacket traveling along the short arm (Fig. 2.7b). After the second beamsplitter, the wavepacket originally occupying two time bins now occupies three, where only the wavepacket peak at the center of each frame interferes constructively (destructively) for the + (−) port. The earliest and the latest wavepacket peaks of the state do not interfere at the second beamsplitter and hence do not directly give information about the phase state. The situation is reversed when the state $|f_1\rangle$ is incident on the interferometer (not shown).

The non-interfering events that are detected in the first and third time bins are classified as inconclusive events and are discarded at the end of communication session. Overall, half of the phase states that propagate through the interferometer are observed in the first or the third time bin. This is precisely due to the probabilistic nature of the first beamsplitter, which sometimes causes both the peaks of the wavepacket to propagate through the same path (shorter or longer arm). Therefore, the efficiency of a phase basis measurement with an interferometer is only 50%. This is further discussed in the later chapters for higher dimensional protocols and a detailed discussion on the efficiency of the interferometric method is provided in Appendix A.

The output of the interferometers is coupled into single-photon counting modules that detect the photon and give an output signal pulse. The output of the single-photon counting modules can be time-tagged with respect to a global synchronization clock that is publicly shared from Alice's FPGA to Bob's time-to-digital converter.

2.5 Limitations of $d = 2$ QKD Systems

Most current QKD systems are qubit-based ($d = 2$) variants of this time-phase protocol, where only one secret bit of information is impressed per photon using various degrees-of-freedom, such as polarization, phase, orbital angular momentum, or time.

For QKD systems operating over relatively short distances where the channel loss is low, the secret key rate is limited by the saturation of single-photon detectors. Detector saturation arises from the so-called detector deadtime, the time over which the detector is not responsive to a photon after a detection event. In greater detail, consider a QKD system where Alice encodes information on photonic wavepackets every period of the system master clock (every time slot). To avoid missing photon detection events, the clock period needs to be longer than the detector deadtime, thus limiting the overall system photon rate and the secure key rate. The advent of cryogenic-based superconducting single-photon nanowire detectors has significantly improved the detector recovery time, but the detected photon rate is still limited to 10–100 MHz [25]. In comparison, the temporal window required to create a photonic state can be less than 1 ps so that the number of photonic states that can be generated is $\sim 10^4$ times greater than can be detected directly.

One solution to this problem is to use the high-dimensional temporal degree-of-freedom for encoding information. Here, the photon is placed in a single time bin within a window (frame) of d contiguous time bins, and the mean photon number is adjusted so that there is approximately one photon per frame. The frame and time-bin size is then adjusted to just reach detector saturation, allowing the extraction of $\log_2 d$ bits of information per detected photon in comparison with other protocols operating at the detector saturation limit that attempt to fill every time slot. Another advantage of the temporal degree-of-freedom is that many standard classical communication components can be used in the QKD system.

In the opposite limit where the quantum channel loss is large, high-dimensional protocols also have an advantage. In particular, the extractable secure key rate can be larger for a high-dimensional protocol as compared to a qubit protocol [26]. In fact, at large errors, there may be no extractable secure key for qubit protocols, whereas some key can be obtained from the higher-dimension approaches. In Chap. 3, I discuss the implementation of a $d = 4$ time-phase QKD system that alleviates some of the practical challenges of $d = 2$ QKD protocols.

Another solution to alleviate the detector saturation problem is to multiplex independent signals using various multiplexing techniques. The simplest and obvious choice is the wavelength-division multiplexing (WDM) technique which utilizes light sources of different wavelengths, typically in a broad spectrum such as the C- or L-telecommunication band, to transmit a large amount of information using a single fiber [27]. Each signal stream is individually encoded using a separate transmitter, and then a multiplexer is used to combine the optical signals at different wavelengths into a fiber. The spacing between the spectral bands used for communication must be greater than the spectral width of the pulses. This requirement is easily satisfied if the linewidth of the lasers is narrow [27, 28]. At the receiver, an optical demultiplexer, which typically consists of optical filters, is used to separate the signals into different wavelengths and then couple them into separate detectors.

Typically, the time-delay interferometers that are used to measure the phase basis states have an operating wavelength range that covers the entire C-band (1530–1565 nm), which means just one interferometer can be used to measure many WDM channels. Thus, the secret key rate of the system can be multiplied by several factors with the same detection setup.

2.6　Conclusion

In this chapter, I discuss the fundamental building blocks of a $d = 2$ time-phase QKD system. Using the idea of MUB states and a simple intercept-and-resend attack strategy, I give an intuitive explanation of the "unconditional" security of QKD systems that originate from fundamental properties of quantum mechanics. I also discuss the main eavesdropping strategies and define the levels of security of QKD protocols. Finally, I describe how a time-phase QKD protocol can be realized in a practical implementation and the limitations of $d = 2$ QKD systems.

References

1. C.H. Bennett, G. Brassard, in *1984 International Conference on Computers, Systems & Signal Processing*, Bangalore (1984), pp. 175–179
2. C.H. Bennett, G. Brassard, ACM Sigact News **20**, 78 (1989)
3. P. Townsend, J. Rarity, P. Tapster, Electron. Lett. **29**, 1291 (1993)
4. P.D. Townsend, Electron. Lett. **30**, 809 (1994)
5. P.D. Townsend, Nature **385**, 47 (1997)
6. F.M. Spedalieri, Opt. Commun. **260**, 340 (2006). http://dx.doi.org/http://dx.doi.org/10.1016/j.optcom.2005.10.001
7. V. Scarani, H. Bechmann-Pasquinucci, N.J. Cerf, M. Dušek, N. Lütkenhaus, M. Peev, Rev. Mod. Phys. **81**, 1301 (2009). http://dx.doi.org/10.1103/RevModPhys.81.1301
8. X. Ma, X. Yuan, Z. Cao, B. Qi, Z. Zhang, npj Quantum Inf. **2**, 16021 (2016)
9. H.E. Guilbert, D.J. Gauthier, IEEE J. Sel. Top. Quantum Electron. **21**, 215 (2015). http://dx.doi.org/10.1109/JSTQE.2014.2375161
10. R. Renner, Int. J. Quantum Inf. **6**, 1 (2008)
11. B. Qi, C.-H.F. Fung, H.-K. Lo, X. Ma, Quantum Inf. Comput. **7**, 73 (2007)
12. M. Lucamarini, I. Choi, M.B. Ward, J.F. Dynes, Z.L. Yuan, A.J. Shields, Phys. Rev. X **5**, 031030 (2015). http://dx.doi.org/10.1103/PhysRevX.5.031030
13. L. Lydersen, C. Wiechers, C. Wittmann, D. Elser, J. Skaar, V. Makarov, Nat. Photonics **4**, 686 (2010)
14. H.-K. Lo, M. Curty, B. Qi, Phys. Rev. Lett. **108**, 130503 (2012). http://dx.doi.org/10.1103/PhysRevLett.108.130503
15. F. Xu, M. Curty, B. Qi, H.-K. Lo, New J. Phys. **15**, 113007 (2013). http://stacks.iop.org/1367-2630/15/i=11/a=113007
16. M. Curty, F. Xu, W. Cui, C.C.W. Lim, K. Tamaki, H.-K. Lo, Nat. Commun. **5**, 1 (2014). https://www.nature.com/articles/ncomms4732
17. H.-L. Yin, T.-Y. Chen, Z.-W. Yu, H. Liu, L.-X. You, Y.-H. Zhou, S.-J. Chen, Y. Mao, M.-Q. Huang, W.-J. Zhang, H. Chen, M.J. Li, D. Nolan, F. Zhou, X. Jiang, Z. Wang, Q. Zhang, X.-B. Wang, J.-W. Pan, Phys. Rev. Lett. **117**, 190501 (2016). http://dx.doi.org/10.1103/PhysRevLett.117.190501
18. Y. Liu, T.-Y. Chen, L.-J. Wang, H. Liang, G.-L. Shentu, J. Wang, K. Cui, H.-L. Yin, N.-L. Liu, L. Li, X. Ma, J.S. Pelc, M.M. Fejer, C.-Z. Peng, Q. Zhang, J.-W. Pan, Phys. Rev. Lett. **111**, 130502 (2013). http://dx.doi.org/10.1103/PhysRevLett.111.130502
19. A. Rubenok, J.A. Slater, P. Chan, I. Lucio-Martinez, W. Tittel, Phys. Rev. Lett. **111**, 130501 (2013). http://dx.doi.org/10.1103/PhysRevLett.111.130501
20. R. Valivarthi, Q. Zhou, C. John, F. Marsili, V.B. Verma, M.D. Shaw, S.W. Nam, D. Oblak, W. Tittel, Quantum Sci. Tech. **2**, 04LT01 (2017). http://stacks.iop.org/2058-9565/2/i=4/a=04LT01
21. H.-K. Lo, X. Ma, K. Chen, Phys. Rev. Lett. **94**, 230504 (2005). http://dx.doi.org/10.1103/PhysRevLett.94.230504

22. C.C.W. Lim, M. Curty, N. Walenta, F. Xu, H. Zbinden, Phys. Rev. A **89**, 022307 (2014). http://dx.doi.org/10.1103/PhysRevA.89.022307
23. Z. Cao, Z. Zhang, H.-K. Lo, X. Ma, New J. Phys. **17**, 053014 (2015). http://stacks.iop.org/1367-2630/17/i=5/a=053014
24. A.V. Gleim, V.I. Egorov, Y.V. Nazarov, S.V. Smirnov, V.V. Chistyakov, O.I. Bannik, A.A. Anisimov, S.M. Kynev, A.E. Ivanova, R.J. Collins, S.A. Kozlov, G.S. Buller, Opt. Express **24**, 2619 (2016). http://dx.doi.org/10.1364/OE.24.002619
25. F. Marsili, V.B. Verma, J.A. Stern, S. Harrington, A.E. Lita, T. Gerrits, I. Vayshenker, B. Baek, M.D. Shaw, R.P. Mirin et al., Nat. Photonics **7**, 210 (2013)
26. J. Leach, E. Bolduc, D.J. Gauthier, R.W. Boyd, Phys. Rev. A **85**, 060304 (2012). http://dx.doi.org/10.1103/PhysRevA.85.060304
27. B. Saleh, M. Teich, *Fundamentals of Photonics*. Wiley Series in Pure and Applied Optics (Wiley, Hoboken, 2007). https://books.google.com/books?id=Ve8eAQAAIAAJ
28. D. Hillerkuss, R. Schmogrow, T. Schellinger, M. Jordan, M. Winter, G. Huber, T. Vallaitis, R. Bonk, P. Kleinow, F. Frey et al., Nat. Photonics **5**, 364 (2011)

Chapter 3
High-Dimensional Time-Phase QKD

Since the demonstration of the first prototype QKD system [1], the field of quantum encryption has progressed rapidly in the following three decades. In most conventional QKD systems, the information is encoded in two-dimensional (qubit) states of a photon. Despite the remarkable progress in experimental realizations of QKD protocols, the secret key rates of qubit-based QKD systems are significantly lower than the rates that can be achieved with classical communication systems [2]. The secret key rates of the qubit-based systems are ultimately constrained by experimental non-idealities, such as the rate at which the quantum states can be prepared, or the long recovery time of the single-photon detectors. One solution to overcome these physical and practical limitations is to encode information in high-dimensional quantum states ($d > 2$), where $\log_2 d$ bits of information can be encoded on each photon [3–6]. High-dimensional encoding has two major implications in the overall system performance. First, it can overcome the long recovery time of single-photon counting modules by encoding more bits per transmitted photon. Second, due to the larger state space, high-dimensional quantum states are more resistant to noise in the quantum channel, which means protocols implemented with these states can tolerate a higher quantum bit error rate than most qubit-based protocols [5].

In high-dimensional schemes, information can be encoded using various degrees-of-freedom, such as time-phase (or frequency) [7–12], spatial modes [13–16], or a combination of these modes [17]. Here, I consider a high-dimensional encoding scheme, in which the information is encoded in frames of time bins, and the presence of Eve is monitored by transmitting states that are discrete Fourier transform of the time-bin states, also known as the phase, or frequency states [7]. The Fourier transform states are a high-dimensional analog of the $d = 2$ phase states discussed in Chap. 2. These states are mutually unbiased with respect to time, which means when a state is prepared in one basis and measured in the other, the measurement results in a uniformly random outcome.

© Springer Nature Switzerland AG 2018
N. T. Islam, *High-Rate, High-Dimensional Quantum Key Distribution Systems*,
Springer Theses, https://doi.org/10.1007/978-3-319-98929-7_3

One major challenge of implementing high-dimensional time-bin QKD protocols is the measurement of the phase states. Past efforts have required intricate measurement schemes, such as using fiber Bragg gratings (FBGs) [9, 18, 19], where the phase states are created by passing a single-photon wavepacket through an FBG that chirps the wavepacket in the time-frequency domains. When the photon arrives at the receiver, Bob decodes the quantum state using a conjugate dispersion FBG coupled into a single-photon detector. While such a scheme is capable of measuring the phase states, matching the dispersion of FBGs at the transmitter and receiver is experimentally challenging, mainly due to environmental disturbances.

Another proposed method to detect the phase states is to use a network of $d - 1$ unequal path length (time-delay) interferometers arranged in a tree-like configuration [7]. The primary challenge with this approach is to stabilize the $d - 1$ time-delay interferometers to sub-wavelength distance scales over a long period of time (minutes to hours). Typically, environmental fluctuations in temperature and pressure, and vibrations cause the interferometers to drift, which makes it difficult to transmit information for a long period of time. To eliminate the effect due to environmental fluctuations, some of the $d = 2$ QKD systems are implemented with a feedback mechanism to actively stabilize the interferometer during the communication session [20, 21]. However, active stabilization of $d - 1$ interferometers in a d-dimensional system increases receiver complexity. Additionally, active stabilization requires two-way communication between Alice and Bob, which can lead to potential side-channel attacks [22].

A solution to this problem is to use passively stabilized interferometers that have been developed over the last decades for use in phase- and frequency-based classical communication protocols [23, 24]. To address the environmental disturbances, these passive devices use athermal design principles in which materials of various thermal expansion coefficients are combined to achieve thermal compensation [25–27]. In addition, fluctuations due to vibrations and pressure are reduced by using a hermetically sealed compact package.

In this chapter, I describe a $d = 4$ time-phase QKD system that is a high-dimensional analog of the protocol discussed in Chap. 2. First, I describe how the quantum states in the time and phase basis are generated and detected in this protocol. Specifically, I describe the measurement device required to detect the $d = 4$ phase basis states, which consists of three time-delay interferometers arranged in a tree-like configuration. I show that the experimentally generated quantum states have high fidelity, and are near-perfect MUB states. In addition, I discuss the security of this protocol using entropic-uncertainty principle and finite-key estimates, which is developed in collaboration with Dr. Charles Lim from the National University of Singapore. Using the tight security bounds and novel experimental techniques, I show that the system can generate high secret key rates. Finally, I conclude the discussion with some potential improvements that can simplify the experiment.[1]

---

[1]The main results of this chapter are published in Refs. [28] and [29].

3.1 Time-Phase QKD Protocol

The d-dimensional time basis states are denoted by $|t_n\rangle$ $(n = 0, \ldots, d-1)$, in which a photonic wavepacket of width Δt is localized to a time bin n of width τ within a frame of d contiguous time bins. The time-basis states for $d = 4$ are illustrated in the left panel of Fig. 3.1.

To monitor the presence of an eavesdropper, I use d-dimensional phase states. They are a linear superposition of all the time states with distinct phase values determined by the discrete Fourier transformation

$$|f_n\rangle = \frac{1}{\sqrt{d}} \sum_{m=0}^{d-1} \exp\left(\frac{2\pi i n m}{d}\right) |t_m\rangle \quad n = 0, \ldots, d-1 \tag{3.1}$$

and illustrated in the right panel of Fig. 3.1 for the specific case of $d = 4$. The phase states have a multi-peaked frequency spectrum with a peak spacing $1/\tau$ and width $\sim 1/2\Delta t$. The carrier frequency of each phase state is shifted with respect to the others. The phase states are also mutually unbiased with respect to the time-basis states, that is, $|\langle t_n | f_m \rangle|^2 = 1/d$.

When the quantum states arrive at Bob's receiver, a beamsplitter is used to randomly direct the incoming quantum photonic wavepackets to either a time or phase measurement device. The time-basis states can be measured with a single-photon counting detector connected to a high-speed time-to-digital converter. The measurement of the phase-basis states requires three delay interferometers (DI) arranged in a tree-like configuration as shown in Fig. 3.2a.

The phase states in $d = 4$ have four contiguous time bins occupied by wavepacket peaks of different relative phases. When the state $|f_0\rangle$ is incident on the first interferometer, as illustrated in Fig. 3.2c, the part of the wavepacket that propagates through the longer arm of the interferometer is shifted temporally by 2τ, relative to the part that takes the shorter arm. When both the parts arrive

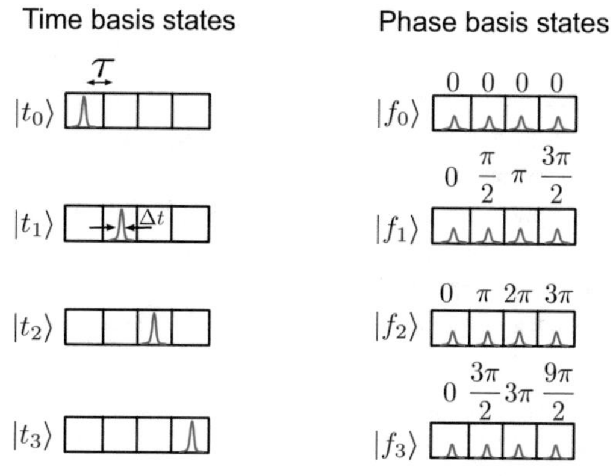

Fig. 3.1 Time and phase basis states in $d = 4$. The phase basis states (right panel) are discrete Fourier transform of the time basis states (left panel) with the phases determined by the unit coefficient in Eq. (3.1)

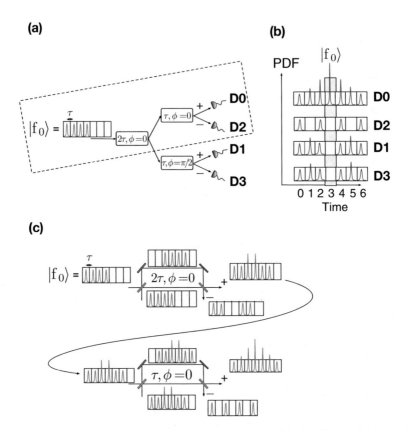

Fig. 3.2 The phase-basis measurement scheme. (**a**) Illustration of the interferometric tree used to measure the phase states. (**b**) Expected photon probability distribution (PDF) at the output of the interferometers when the phase state $|f_0\rangle$ is injected into the system. (**c**) Illustration of how the phase state $|f_0\rangle$ propagates through the top branch of the interferometric tree. Adapted from Islam et al. [29], Fig. 2

at the second beamsplitter, the peaks occupying the central time bins interfere constructively (destructively) at the $+$ $(-)$ port of the interferometer. The portion of the wavepackets that occupy the outer time bins do not interfere. The two output ports of the interferometer are coupled into a second set of interferometers in the next layer of the interferometric tree. Here, I only describe the interference of the wavepacket that propagates through the upper branch of the interferometric tree, indicated by the dashed box in Fig. 3.2a. A similar analysis can also be performed for the lower branch of the tree.

When the wavepacket propagates through the second interferometer (lower panel of Fig. 3.2c) with a time-delay τ and phase $\phi = 0$, at the output ports, all 7 time bins are occupied. The highest (lowest) probability of photon detection occurs in the central time bin of the $+(-)$ output port. This is due to the constructive and destructive interference of all four wavepacket peaks in the initial state $|f_0\rangle$, and therefore, provides a direct means of detecting the phase state. The other occupied time bins, except the outer most ones, reveal information about interference of a subset of the incident wavepacket peaks. Although the measurement of these off-centered peaks can provide partial information about the coherence of the wavepacket, complete knowledge of the incoming state can only be obtained from the events that occur in the central time-bin. A detailed investigation reveals that, for a QKD demonstration, the measurement of these time bins (4–6) is not necessary. Additionally, the error rate in the phase basis does not increase due to the spillover of the wavepacket into the neighboring (adjacent) time bins, since only the events in the central time bins are considered to be conclusive and used in the post-processing step. The expected interference patterns when state $|f_0\rangle$ propagates through the interferometric setup is shown in Fig. 3.2b.

A similar analysis shows that the central time-bin of each output port of the interferometric tree is directly related to a specific phase state, that is, there is a one-to-one mapping between the input phase state and the output detector in which it is observed. Specifically, constructive interference occurs in the output detector Dn when the state $|f_n\rangle$ is incident on the cascade, and destructive interference is observed in the remaining three ports. Such a one-to-one mapping of the input phase state and the output in which it is observed is not unique to only $d = 4$; the procedure can be extended to an arbitrary d using $2d - 1$ time-delay interferometers as shown in Ref. [7].

Although the interferometric tree can be extended to measure any arbitrary dimensional phase states, the probability of conclusively determining a phase state using $2d - 1$ time-delay interferometers is only $1/d$. This is due to the spillover of the wavepacket to the neighboring time bins as observed above for $d = 4$ and in Chap. 2 for $d = 2$. If the phase states are used to determine the presence of an eavesdropper, instead of generating the secret key, the inefficiency of the measurement scheme does not affect the secret key rate. Instead, the inefficiency only increases the duration of time required to obtain a large enough sample size to accurately determine the presence of Eve. This can be overcome using a novel two-photon interference based measurement scheme discussed in Chap. 5.

3.2 Experimental Details

An illustration of the experimental system used to implement this protocol is shown in Fig. 3.3. The QKD system is based on a prepare-and-measure scheme, in which Alice prepares the quantum states in one of the two randomly chosen bases, and Bob measures the quantum states in a randomly chosen basis.

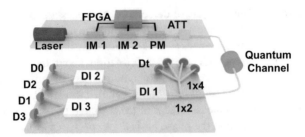

Fig. 3.3 Schematic of the experimental setup. Alice modulates the intensity and phase of a cw laser to create the $d = 4$ time and phase basis states. She then transmits the quantum states via an untrusted quantum channel to Bob. Upon arrival, Bob directs the incoming photon using a 90/10 coupler to time and phase basis measurement devices, respectively. Adapted from Islam et al. [28, e1701491, Fig. 1]

3.2.1 Transmitter

At Alice's transmitter, the quantum photonic states (signal and decoys) are created by modulating a 1550 nm frequency-stabilized continuous laser (Wavelength Reference, Clarity-NLL-1550-HP) using electro-optic intensity and phase modulators (all intensity and phase modulators are from EOSpace).

The entire system is controlled by independent serial pattern generators realized with a high-speed transceiver on an FPGA (Altera Stratix V). The clock frequency of the FPGA is set to 10 GHz. Specifically, a 5 GHz sine-wave generator phase locked to the FPGA is used to drive an intensity modulator (not shown), which is biased at the fringe minimum. The amplitude of the sine-wave generator is adjusted to twice the V_π voltage of the intensity modulator, which creates a periodic train of ~66-ps-width (FWHM) pulses. The short optical pulses are useful for reducing the total system jitter, which in turn decreases the quantum bit error rate of the system. These pulses then pass through an intensity modulator (IM 1), which is driven with an FPGA-based pattern generator to curve the pulse train and generate the data pattern for either the time or phase basis states. This intensity modulator has a high extinction ratio (>40 dB), which is important for suppressing any leakage of light when the modulation is off. As mentioned in Chap. 2, a high-extinction ratio intensity modulator is critical for time-bin encoding schemes to ensure that uncorrelated background events do not result in erroneous bit values. A second intensity modulator (IM 2) of extinction ratio ~30 dB, driven with a combination of two independent FPGA signals, adjusts the amplitude of the optical wavepackets to generate the phase states and the decoy states. In this experiment, the time and phase basis states are generated with three different mean intensities, corresponding to mean photon numbers μ_1, μ_2, and μ_3. The quantum states with the highest mean photon number (μ_1) are referred to as signal states, and the ones corresponding to mean photon numbers μ_2 and μ_3 are referred to as decoy states. Finally, three independent FPGA signals are combined in a 3×1 coupler (Marki Microwave

PD30R412), and the output signal is used to drive a phase modulator, which applies a phase to the optical wavepacket proportional to the input signal. This means, when three independent signals are combined, three distinct phase values can be imposed in any given time bin.

Overall, the rate of state preparation is set to 625 MHz, and each time-bin width is set to 400 ps. The time-bin width is chosen to be 400 ps so that, when the total system jitter is taken into account, the error rate in the time basis is ~2%. In principle, the quantum bit error rate can be decreased further by increasing the time-bin width. However, this also decreases the state preparation rate ($1/d\tau$), and therefore the secret key generation rate.

The time- and phase-basis are chosen with probabilities $p_T = 0.90$ and $p_F = 0.10$, respectively, so that a large fraction of the detected events is used for secret key generation, and only a small fraction is used to monitor the presence of an eavesdropper. In principle, the optimum probabilities of transmitting the time and phase basis, which maximize the secret key rates, can be determined by simulating the secret key rate for every combination of probabilities at every channel loss (see Sect. 3.8). However, in the experiment, using different probabilities at every quantum channel loss is difficult, since this requires changing both the transmitter and receiver setup.

Calibration of Mean Photon Numbers

After the intensity and phase modulation, the classical photonic wavepackets are attenuated to the single-photon level using a variable optical attenuator (ATT). To calibrate the mean photon number, a 99/1 beamsplitter is placed after the variable optical attenuator (not shown in Fig. 3.3 for clarity). The output of the beamsplitter corresponding to the 99% transmission is input into a power meter, and the other fraction corresponding to the 1% transmission is sent to Bob via an untrusted quantum channel.

The mean photon numbers are then estimated based on the reference power observed in the power meter, the power ratio of the signal to decoy states, as well as the probability of transmitting each of these. Specifically, the wavelength of the laser is well known, and therefore the average number of photons transmitted per second to Bob can be estimated as

$$\bar{\mu} = \frac{\alpha P_{\text{ref}}}{hc/\lambda}, \tag{3.2}$$

where $\alpha = 1/99$ is the scaling factor that accounts for the 99/1 ratio of the measured power to the transmitted power; h is the Planck's constant; c is the speed of light; λ is the wavelength of Alice's source. In order to estimate the mean photon number in the signal state, it is necessary to relate $\bar{\mu}$ to the probabilities of transmitting signal and decoys, as well as the power ratio of the signal to decoys. The probabilities of transmitting signal and decoy states are stored in the FPGA memory. The ratio of the signal power to the decoy power is estimated using a classical photo-receiver before the attenuation. This is also cross-checked with a single-photon detector in the single-photon regime. Specifically, a fixed pattern of the highly attenuated signal and decoy signals is detected using a single-photon detector, and the detector pulses are time-tagged using a timing analyzer (PicoHarp 300 from PicoQuant) to histogram the time-of-arrival of the states. The area under the histogram peaks corresponds to the total number of photons, and thus proportional to the intensity. The amplitude of the signal to decoys can be estimated and adjusted based on these timing histograms.

Based on these parameters, the weighted average of the photon number can be written as

$$\bar{\mu} = \mu_1 P_{\mu_1} + \mu_2 P_{\mu_2} + \mu_3 P_{\mu_3}, \tag{3.3}$$

where P_{μ_1}, P_{μ_2}, and P_{μ_3} are the probabilities of transmitting states corresponding to mean photon numbers μ_1, μ_2, and μ_3. To estimate the mean photon number μ_1, I divide both sides of Eq. (3.3) by μ_1

$$\frac{\bar{\mu}}{\mu_1} = P_{\mu_1} + \frac{\mu_2}{\mu_1} P_{\mu_2} + \frac{\mu_3}{\mu_1} P_{\mu_3}, \tag{3.4}$$

where the ratios of the mean photon numbers on the right-hand side are determined during calibration using the classical photo-receivers, as well as from the timing histograms of the single-photon detector pulses as discussed above. Therefore, the mean photon number per signal state can be obtained by solving for μ_1.

3.2.2   Quantum Channel

In this experiment, the quantum channel is simulated using a second variable optical attenuator which has the advantage that the channel loss can be continuously tuned to simulate any length of fiber or free-space links. However, there are additional challenges that may need to be addressed in a deployed fiber or free-space field test. For example, in the fiber-based channels, the quantum channel loss may be more than what is estimated from the coefficient of loss of a single-mode fiber. This may happen due to additional splicing, coupling and bend in the fiber [18]. Secondly, environmental fluctuations such as variation in temperature may cause the length of the fiber to drift, and hence the timing of the system may need to be synchronized periodically. This can be performed by adjusting the offset of the pulse train with respect to the master clock on the FPGA, and then fine-tuning the temporal alignment using a pico-second precision optical delay line. Finally, the dispersion coefficient of the optical fiber may cause temporal wavepackets to chirp. However, because I create 66-ps-width temporal wavepackets in a 400 ps time bin, the dispersion might have an insignificant affect on the system performance.

3.2.3 Receiver

When the signals arrive at Bob's receiver, he uses single-photon detectors to measure the quantum states in the time basis, or a network of time-delay interferometers coupled to single-photon detectors to measure the quantum states in the phase basis. The incoming signals are split using a 90/10 beamsplitter to direct 90% of the states to the time basis measurement system and 10% to the phase basis system. This choice is made based on the probabilities of transmitting time and phase basis as discussed above.

Superconducting Nanowire Single-Photon Detectors

For both measurement bases, I use commercially available superconducting nanowire single-photon detectors (SNSPDs, Quantum Opus), and the detection events are recorded with a 50-ps-resolution time-to-digital converter (Agilent, Acqiris U1051A), which is synchronized with Alice's clock over a public channel. The SNSPDs used in this experiment have high-detection efficiencies ($>70\%$) up to a count rate of 2 Mcps and low timing-jitter (<50-ps) [30]. Although the dark count rates of these detectors are specified to be <100 cps at 1550 nm, I find that under typical operating conditions, the dark counts can be as high as 300–500 cps. High dark count rates are not a limiting factor for QKD systems operating in low- to moderately high-loss quantum channels, but may become significant if the quantum channel loss is high.

One limiting factor of the detectors used in this experiment is that the efficiency of the detectors starts to drop when the incoming photon rate exceeds 2 Mcps due to the finite reset time (discussed below). To overcome this detector saturation effect, the time-basis measurement device in Bob's receiver system is designed using a multiplexed detection scheme, where a 1:4 coupler is used to direct the quantum states to four SNSPDs placed in parallel, as shown in Fig. 3.3. In this scheme, when two photons arrive at the receiver faster than the deadtime of the SNSPDs, there is a 75% chance that the second photon will go to a different detector, which may potentially be active. This multiplexed detection scheme, along with the greater information content of high-dimensional quantum states, allows the QKD system to generate a high secret-key throughput.

Time-Delay Interferometers

A novel feature of this QKD system is the phase-state measurement device [7, 29] as shown in Fig. 3.3. The first stage delay interferometer (DI 1) has a time-delay of 800 ps, corresponding to a free-spectral range (FSR) of 1.25 GHz and matched to twice the time-bin width of the optical wavepackets. The second stage interferometers (DI 2 and DI 3) have a time-delay of 400 ps. The phases of DI 1 and DI 2 are

Fig. 3.4 Design of Kylia interferometers. Illustration of the internal components of a folded Michelson-type delay-line interferometer. Adapted from Islam et al. [29], Fig. 3

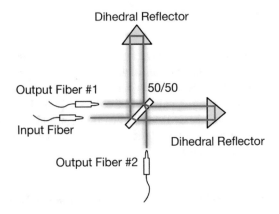

set to 0 radian, and DI 3 is set to $\pi/2$ radian. The phase-state measurement device is built using commercially available time-delay interferometers that are designed to be field-deployable, and hence do not require active path-length stabilization. The stability analysis of the interferometers is presented in Appendix B. Here, I briefly summarize the key findings of the stability analysis.

The delay interferometers used in this experiment are of Michelson-type (Fig. 3.4) with a delay in one arm of the interferometers [25, 27]. In this arrangement, the incoming beam of light is split into two unequal paths using a 50/50 beamsplitter, thus creating an optical path length difference. After reflecting back from the dihedral reflectors, which causes a spatial displacement, the two beams are recombined at the same beamsplitter and are separated into two output ports. This folded design has the advantage that the interferometers can be compact even when the time-delay is relatively large (nanoseconds). Another advantage of this design is that the power difference in two arms is minimal due to identical reflectors in both the arms, which enables high-visibility interference. In addition, the time-delay between the arms can be made arbitrarily small, thus large FSR devices can be fabricated in a compact package. These devices are designed to operate over the classical optical telecommunication C-band and I evaluate their performance at 1550 nm near the middle of the C-band, where the QKD system is operated.

For these devices to be useful either in classical or quantum communication, the overall phase ϕ of the time-delay interferometers must be tunable to high precision, which is realized by changing the optical path of one arm of the interferometer relative to the other [26]. In the Kylia devices, a resistive heater (actuator) is connected to one of the dihedral reflectors, which changes the relative path when a voltage is applied; applying ≤ 3 V results in a path length change of approximately one FSR.

The stability of these interferometers against temperature changes in the environment depends largely on the method for thermal compensation and environmental isolation. While the Kylia design is proprietary, typical athermal delay-line interferometers use materials with a low coefficient of thermal expansion, such as zerodur or ultra low expansion (ULE) glass as a spacer between the mirrors and beamsplitter with a tiny air gap in one path to make the final adjustments of the FSR [25–27]. The interferometers used for this study are constructed of optical components with ULE substrates mounted on a ULE base plate and packaged inside a hermetically sealed aluminum housing, which stabilizes them against temperature and pressure changes in the environment, respectively (L. Fulop, Kylia, 10 Rue deMontmorency, 75003 Paris, France, 2015, personal communication).

The key performance metrics that determine the suitability of these interferometers for QKD experiments are the stability of these devices against environmental changes in a nominally constant temperature and the temperature-dependent phase shift (TDPS). From the characterization measurements discussed in Appendix B, I find that the long-term stability of the Kylia devices are better than 3 nm over an hour time-scale at a nominally constant room temperature. Additionally, the TDPS of these devices vary from device-to-device. The 800 ps delay interferometer has a TDPS of 50 ± 17 nm/°C at 37.1 °C, and one of the 400 ps interferometer has a TDPS of 26 ± 9 nm/°C over [22, 50] °C. Given the stability of these devices at nominally constant temperature, I find that the visibility of the interferometers varies less than 1% over an hour time-scale, which corresponds to <0.5% variation in the quantum bit error of the system.

3.3 System Characterization

The most important figure-of-merit for any QKD system is the secret key rate, which depends on several important criteria, such as the quality and rate of state preparation, the maximum rate of state detection, and the quantum bit error rates in both the bases. Below I characterize some of the key parameters of the system. For these characterization measurements, I fix the attenuation in the quantum channel to 14 dB, equivalent to a fiber length of 70 km, assuming a coefficient of loss of 0.2 dB/km at 1550 nm wavelength. The mean photon detection rate at this loss is 5.88 ± 0.13 Mcps.

3.3.1 Quality of State Preparation and Detection

I characterize the quality of state preparation and detection by measuring each of the eight quantum states in both the bases. The measurement outcomes are then used to determine the probability-of-detection matrix, in which the elements represent the overlap of the quantum states such as $|\langle f_n | t_m \rangle|^2$, $|\langle f_n | f_m \rangle|^2$, $|\langle t_n | t_m \rangle|^2$, and $|\langle t_n | f_m \rangle|^2$, $\{m, n\} \in \{0, 1, 2, 3\}$. These overlap elements can be used to determine the quality of MUB, as well as the quantum bit error rates of the system.

Ideally, when a quantum state is prepared in one basis and measured in the other, the measurement results in a uniformly random outcome: $|\langle t_n | f_m \rangle|^2 = 1/d$ or

$|\langle f_n|t_m\rangle|^2 = 1/d$. For example, when a time basis state is measured in the phase basis, it has an equal probability of being detected as any of the phase states. This is illustrated graphically in Fig. 3.5a for the ideal case, where the state preparation and detection are assumed to be perfect. The matrix elements in the first (third) quadrant of Fig. 3.5a represent the cases where time (phase) basis states are measured in the phase (time) basis. In this case, the probability that a time (phase) basis state is detected as any of the phase (time) basis states is 0.25.

On the other hand, the elements in the second and fourth quadrants represent the cases where states are measured in the same basis in which they are prepared. In the ideal scenario, the probability of detecting the correct state is 1, as represented by the strong correlation of the diagonal elements in Fig. 3.5a. The off-diagonal elements in these quadrants represent the probability of detecting an incorrect state. For a given input state, the sum of all off-diagonal elements represents the probability of a detecting a quantum bit error. For ideal state preparation and detection, this sum is zero.

In Fig. 3.5b, I plot the experimentally determined values of the overlap elements. The raw values of the matrix elements are provided in Fig. 3.5c. The statistical uncertainties for these measurements are between 2% and 4%, which is reasonable given that the state preparation and detection in an experimental measurement is not perfect. The elements in the first and third quadrants are very close to the ideal value of 0.25, with a statistical standard deviation of 0.006 and 0.02, respectively. The elements along the diagonal indicate a strong correlation, as expected when states are prepared and measured in the same basis. The off-diagonal elements in these quadrants represent the quantum bit errors as discussed in the next subsection.

3.3.2   Error Rates and Visibility

The quantum bit error rates are critical system parameters that determine the number of bits that need to be revealed during error correction, and are therefore a limiting factor in the secret key generation of any QKD system. These errors may occur due to experimental flaws, such as optical misalignment, uncorrelated photons entering the quantum channels, dark counts, noise from the readout electronics of the detector, leakage of background light through the intensity modulators, the total system jitter, among others. Additionally, the quantum bit errors may also occur due to imperfect state preparation and visibility of the time-delay interferometers.

During the 100 s of data collection, the total number of observed events in the time and phase basis are determined to be 4.94×10^8 and 1.63×10^6, respectively, and the corresponding quantum bit error rates are measured to be $(1.94 \pm 0.17)\%$ and $(3.69 \pm 0.04)\%$. Note that the quantum bit error rates for this particular measurement are smaller than the values in Fig. 3.5c. This is because the two data sets were taken at different times, and in this case, the system was better aligned and optimized.

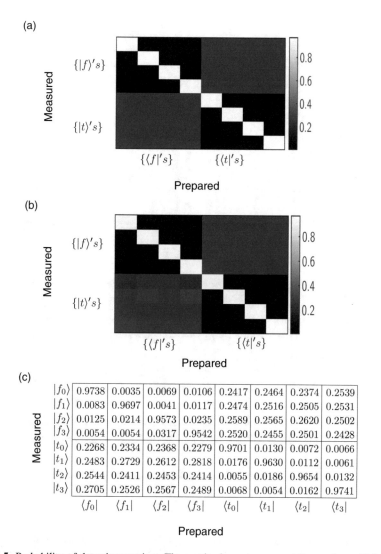

Fig. 3.5 Probability-of-detection matrices. The matrix elements represent the overlaps of (**a**) ideal quantum states and (**b**) the experimentally prepared and measured quantum states. (**c**) The raw values of the experimentally determined overlap elements in the time and phase basis. Adapted from Islam et al. [29], Fig. 11

In Fig. 3.6a, I plot the timing histograms of the detected events when each of the four time-basis states is measured using an SNSPD. The FWHM of the timing histograms are measured to be $\sim$110 ps, which is small compared to the time-bin width of 400 ps. Therefore, it can be concluded that the system timing jitter is not the primary source of error in the QKD system. A detailed investigation shows that the main contributors to the quantum bit error rates in this QKD system are the

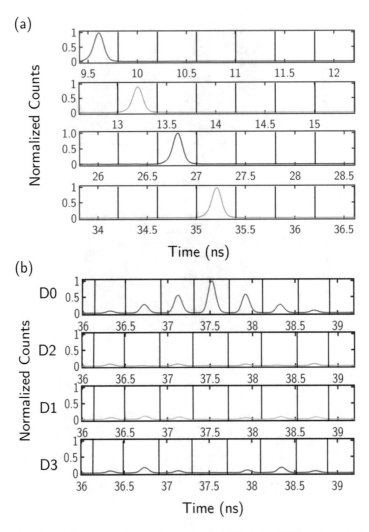

Fig. 3.6 Experimentally measured time-phase states. (**a**) The timing histograms of the temporal states $|t_0\rangle$, $|t_1\rangle$, $|t_2\rangle$, and $|t_3\rangle$ as measured using an SNSPD. (**b**) The timing histograms of the phase state $|f_0\rangle$ as measured at the output of the interferometric setup using four SNSPDs. Adapted from Islam et al. [29], Fig. 12

leakage of light from the intensity modulator, and the spurious events from the readout electronics of the single-photon detectors (Sect. 3.9). Approximately, 1% of the errors can be attributed to these factors.

In Fig. 3.6b, I plot the timing histograms when the phase state $|f_0\rangle$ propagates through the interferometric setup, and the photons are detected in detectors D0, D2, D1, D3 (top to bottom panels). The observed interference pattern is similar to the expected intensity pattern as shown in Fig. 3.2b. To determine the stability of the

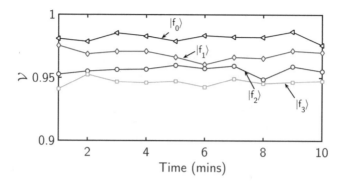

Fig. 3.7 State-dependent visibility. Visibilities of all phase states as a function of time. Adapted from Islam et al. [29], Fig. 13

interferometric setup, in Fig. 3.7, I plot the visibility, $\mathcal{V}$, of all four phase states as a function of time during which the data was collected. The visibility is determined by

$$\mathcal{V} = \frac{\mathcal{P}_+ - \mathcal{P}_-}{\mathcal{P}_+ + \mathcal{P}_-}, \tag{3.5}$$

where $\mathcal{P}_+$ is the probability of detecting the photon in the central bin of the expected bright port n, and $\mathcal{P}_-$ is the probability of finding the photon in the central bin of any of the other ports. In Fig. 3.7, I subtract 1.94% of the events to correct for the effect of leakage, detector noise, etc., and to estimate the imperfect visibility that arises from the state preparation and measurement flaws.

Figure 3.7 shows that the visibilities of all the phase states are approximately constant during the 10 min of data collection. The visibility of the phase state $|f_0\rangle$ is ~98%, which is in agreement with the characterization measurement performed with classical light as demonstrated in Appendix B. On the other hand, the visibilities of the phase states $|f_1\rangle$, $|f_2\rangle$, and $|f_3\rangle$ are smaller. The lower visibility for these measurements can be attributed to the imperfect state preparation.

Specifically, recall from Sect. 3.2 that to impose the distinct phase values on the wavepacket peaks, three independent signals from an FPGA are combined in a 3×1 coupler, and the output signal is used to drive a phase modulator. The phase values of these states have repeated patterns, as shown in Fig. 3.1. For example, consider the phase values on the wavepacket peaks of $|f_1\rangle$; the phase difference between successive peaks increases in steps of $\pi/2$. Similarly, the phase difference between successive peaks of $|f_2\rangle$ and $|f_3\rangle$ increases in steps of π and $3\pi/2$, respectively. To generate these phases, I combine three independent, equal-amplitude signals from the FPGA as shown in Fig. 3.8a. For example, to create the phase state $|f_0\rangle$, the first FPGA signal going into amplifier 1 (Amp1, all amplifiers are from JDSU Model H301) is set to 'on' between time bin 1 and 3. Similarly, the second FPGA signal going into Amp2 is set to 'on' between 2 and 3. Finally, the third FPGA signal going into Amp3 is set to 'on' only in time bin 3. As a result, the combined output

(a)

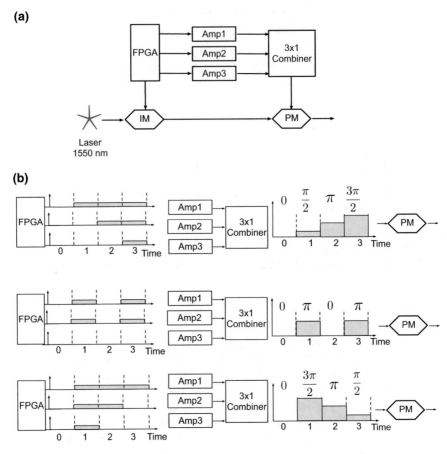

(b)

Fig. 3.8 Schematic of the phase-state generation setup. (**a**) Three independent signals from an FPGA are combined in a 3 × 1 combiner to generate all the phase states. (**b**) A detailed illustration of the signal patterns that are combined to create the phase states $|f_1\rangle$, $|f_2\rangle$, and $|f_3\rangle$. Adapted from Islam et al. [28, e1701491, fig. S6]

of 3 × 1 combiner is a step-like function that matches the pattern of the phase values required to create the phase state $|f_1\rangle$; each step corresponds to a phase shift of $\pi/2$. Figure 3.8b also shows the combinations of signals required to generate the phase states $|f_2\rangle$ and $|f_3\rangle$. Note that the step-like signal required to create the phase values for $|f_3\rangle$ is the inverse of the signal required to create $|f_1\rangle$.

Although this method of state preparation can generate high-fidelity phase states, it has two specific issues that can lead to a slight decrease in the preparation quality as seen from the low visibility of the states $|f_1\rangle$, $|f_2\rangle$, and $|f_3\rangle$ in Fig. 3.7. First, each of the three independent signals needs a separate amplifier, which means the gain of the amplifiers need to be matched very accurately. In principle, these step-like signals can also be generated if the FPGA signals are combined before they are amplified. In that case, only one amplifier is required. However, I find that most commercially available variable gain amplifiers saturate with the combined power of three independent signals. Second, the propagation time of the signals from the FPGA to the coupler can vary, which also results in imperfect phase states. These problems can be addressed in the future experiments by using precision high-speed time-delays and step attenuators.

With the three-amplifier setup, I find that the ability to set the $\pi/2$-phase depends on how precisely the gain of the amplifiers can be adjusted. Once the phase $\pi/2$ is set, the phase of the second and third steps corresponding to π and $3\pi/2$ phase values can be tuned to an accuracy of 0.06 and 0.15 radians, respectively. This results in a slight error in the generation of the phase state $|f_1\rangle$, $|f_2\rangle$, and $|f_3\rangle$ as discussed above.

3.4 Sketch of the Security Proof

The security analysis presented below is performed in collaboration with Dr. Charles Lim. A detailed analysis is provided in the Supplementary Information of Ref. [28]. Here, I provide a sketch of the proof and the final secret key length equation that is used to calculate the finite-key secret key rate of the QKD system.

The security of our QKD system is derived using entropic uncertainty relations, which bounds Bob's correlation with Alice's secret key, and Eve's knowledge of the shared quantum states using an entropic relation [31, 32]. Most of the previous security analyses of high-dimensional QKD systems were either secure against collective attack strategies, and did not consider the effect of imperfect MUB or the finite-key effects [9, 11, 18], or the bounds were applicable only for single-photon states [5]. Unlike the previous analyses, we analyze the security of this protocol in the limit of finite-key length, against general (coherent) attacks and we consider the effect of imperfect state preparation as characterized by the parameter $c := -\log_2 \max_{i,j} |\langle f_i|t_j\rangle|^2$. We also assume that Alice transmits quantum states of three different mean intensities μ_1, μ_2, and μ_3 to implement the decoy-state technique.

The security of this protocol is defined by two criteria. First, the protocol is considered as ϵ_{cor}-correct if the probability that Alice and Bob do not have identical secret key after the error correction step is extremely small, defined by $\Pr[S_A \neq S_B] \leq \epsilon_{\text{cor}}$, where $S_A(S_B)$ is the secret key of Alice (Bob). Second, the protocol is known as ϵ_{sec}-secret if the final secret key shared between Alice and Bob is ϵ_{sec} *close* to a completely random string [31]. The secrecy criterion also guarantees that the secret key is composable, which means that key can be used for any cryptographic tasks, such as one-time pad encryption. Combining these two definitions, we say that our protocol is $\epsilon - secure$ if it is both ϵ_{cor}-correct and ϵ_{sec}-secret, that is $\epsilon \leq \epsilon_{\text{cor}} + \epsilon_{\text{sec}}$.

Using the entropic uncertainty relations, we find that the secret key length ℓ is given by [28]

$$\ell \leq \max_{\beta \geq 0} \lfloor 2\tilde{s}_{T,0} + \tilde{s}_{T,1}[c - H(e_F^U)] - \mathsf{leak}_{EC} + \Delta_{FK} \rfloor, \tag{3.6}$$

where $\tilde{s}_{T,0}$ ($\tilde{s}_{T,1}$) is the number of vacuum (single-photon) detections in the raw key estimated from the bounds calculated using decoy-state analysis. The quantity e_F^U is an upper-bound on the single-photon *phase error rate* which is a function of the observed error rate in the phase basis; $H(x) := -x \log_2 (x/3) - (1-x) \log_2(1-x)$ is the Shannon entropy for $d = 4$; leak_{EC} is the total number of bits revealed during error correction; $\Delta_{FK} := -\log_2 \left(32\beta^{-8}\epsilon_{cor}^{-1} \right)$ is the finite-key contribution. The secret key length is maximized numerically over the free parameter β satisfying $4\epsilon_{cor} + 18\beta \leq \varepsilon$ (see Ref. [28]). Note that the upper bound on the phase error rate is denoted as λ^U in Ref. [28]. In order to calculate the secret key rate, we divide ℓ by the total duration of the communication session, which is 100 s in all the data presented in this chapter.

As mentioned above, we characterize the quality of state preparation using the quantity c, which we estimate from the experimentally determined probability of detection matrix as shown in Fig. 3.9. Since the quantity is defined as the maximum of all overlaps, we take the worst case value and calculate $c = 1.89$. There is no associated measurement uncertainty in c because it is the worst case value. In principle, when the state preparation and detection is perfect, the overlap between time and phase states is expected to be 0.25, which corresponds to $c = 2$. Essentially, the value of c relates the maximum bits of raw information that can be encoded on a photon. Our estimate of c is very conservative because it takes into account both the imperfect state preparation and measurement, although the definition assumes a perfect measurement. In a real experiment, it is not possible to decouple the non-idealities of the measurement device from the preparation device.

Fig. 3.9 Probability of detection when each input state is measured in both bases. The maximum of the off-diagonal elements when the states are prepared and measured in different bases determines the parameter c. Adapted from Islam et al. [28, e1701491, Fig. 2]

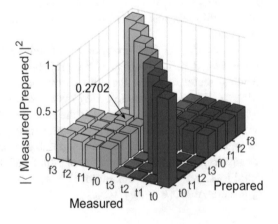

3.5 Demonstration of High-Rate Secret Key Throughput

Combining all the experimental and theoretical tools, I implement the QKD system and transmit secret key over five different quantum channel losses. I calculate the achievable secret key using Eq. (3.6), and plot the lower bound secret key rates as a function of quantum channel loss (fiber length) in Fig. 3.10a. As before, I assume that the coefficient of loss in optical fiber is 0.2 dB/km. For benchmarking this work with previous works, I present a comparison in Table 3.1.

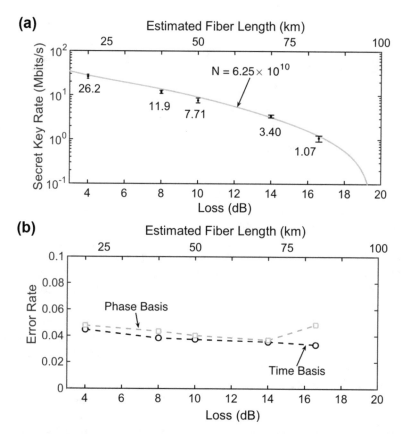

Fig. 3.10 High-rate secure quantum key distribution. (**a**) Experimentally determined secret key rates as a function of the channel loss when Alice transmits $N = 6.25 \times 10^{10}$ quantum states, corresponding to a 100-s-duration communication session. The orange solid line is the simulated secret key rate. For the simulation, the probabilities of transmitting the signal, decoy, and vacuum intensities are set to 0.8, 0.1, 0.1, respectively. The intrinsic error rate in the time basis and phase basis are set to 0.03 and 0.025, respectively. (**b**) Experimentally observed quantum bit error rate in the time and phase basis signal states plotted as a function of channel loss. Adapted from Islam et al. [28, e1701491, Fig. 3]

Table 3.1 Comparison of some notable high-rate QKD systems

	Protocol	Loss (dB)	Equivalent fiber length (km)	Secret key rate (Mbits/s)	Security level
Ref. [33]	T12	7	35	2.20	Collective
		10	50	1.09	
		13	65	0.40	
		16	80	0.12	
Ref. [34]	HD-QKD	0	0	7.0	Collective
		4	20	2.7	
Ref. [18]	HD-QKD	0.1	0.5	23.0	Collective
		7.6	38	5.3	
		12.7	63.5	1.2	
Our work	HD-QKD	4	20	26.2	Coherent
		8	40	11.9	
		10	50	7.71	
		14	70	3.40	
		16.6	83	1.07	

Adapted from Islam [28, e1701491, Table 1]

Overall, the secret key rates observed in this experiment are a few folds improvement over the past results. Specifically, at a channel loss of 4 dB, which corresponds to a 20 km fiber length, the system can generate a secret key rate of 26.2 Mbits/s. This is the highest secret key reported for any QKD systems at this distance. The corresponding error rates in the time and phase basis are 4.5% and 4.8%, respectively, as shown in Fig. 3.10b. The slightly higher error rates at this channel loss is mainly due to the spurious events from the readout electronics of the single-photon detectors that are prevalent at high count rates (see Sect. 3.9). Similar high secret key rates are also observed across other channel losses, extending up to 16.6 dB (83 km fiber length), as illustrated in Table 3.1. The sifted data, experimental parameters, and the statistical uncertainties in the measurement of error rates are provided in Sect. 3.9.

The solid line in Fig. 3.10a represents the simulated secret key rate calculated using parameters that are observed in the experiment. For instance, the simulated curve takes into account the intrinsic error rates, probability of observing dark counts, and the saturation of the single-photon detectors, which are obtained from the experimental data. As a result, I find that the simulation can model the secret key rate very well, within experimental uncertainties, as shown in Fig. 3.10a. The reduced-χ^2 value between the observed and expected data is calculated to be 1.83. The probability of obtaining a reduced-χ^2 value greater than 1.83 for the same degrees-of-freedom is determined to be 13%. A better fit could be obtained if the mean photon numbers between different channel losses were consistent (Sect. 3.9), which is assumed to be the case in the simulation.

From the simulation, it is apparent that the secret key rate drops rapidly beyond a channel loss of 18 dB, which occurs primarily due to the increasing finite-key contribution resulting from the fixed data collection period of 100 s. For a fixed duration of data collection, the total events received in Bob's detectors decrease at higher channel losses, which increases the upper bound phase error rate (e_F^U) of the system.

3.6   Secret Key Rate Simulation

The simulation of the secret key rate requires modeling the quantum channel loss, dark count probability, intrinsic error due to misalignment, the rate-dependent efficiency of the single-photon detectors and all the bounds calculated from the decoy-state technique. In this section, I outline these models, and in later sections I show simulated results when all the parameters are optimized at every channel losses.

Specifically, I assume that the quantum channel is described by a loss of $\eta_{ch} := 10^{-0.2l/20}$, where l is the fiber length, and η_d is the rate-dependent efficiency of the single-photon detectors. The overall loss in the system is given by $\eta_d \eta_{ch}$. Additionally, I assume that the probability of observing a dark count in a given time bin is P_d, and the total number of signals transmitted by Alice is N.

For a given mean photon number μ_k, the total number of detection events in Bob's time basis measurement device can be written as

$$n_{T,k} = p_{\mu_k} p_T^2 N (1 - \exp[-\eta \mu_k] + P_d). \tag{3.7}$$

Similarly, the total number of detection events in the phase basis can be written as

$$n_{F,k} = p_{\mu_k} p_F^2 N (1 - \exp[-\eta \eta_i \mu_k] + P_d), \tag{3.8}$$

where the additional loss factor η_i in the exponent reflects the reduced transmittance due to the insertion loss of the interferometers.

The error rates in the time basis and phase basis are given by

$$m_{T,k} = p_{\mu_k} p_T^2 N \{ e_d (1 - \exp[-\eta \mu_k]) + 0.75 P_d \}, \tag{3.9}$$

and

$$m_{F,k} = p_{\mu_k} p_F^2 N \{ e_d (1 - \exp[-\eta \eta_i \mu_k]) + 0.75 P_d \}, \tag{3.10}$$

respectively, where e_d is the intrinsic error rate which occurs mainly due to the leakage of background light through the intensity modulator. The factor of 0.75 reflects that 75% of the dark count events result in errors.

Finally, to simulate the secret key rate, I calculate the secret key length ℓ from Eq. (3.6). The calculation of ℓ requires using the above equations to determine the bounds of $\tilde{s}_{T,0}$, $\tilde{s}_{F,0}$, $\tilde{s}_{T,1}$, $\tilde{s}_{F,1}$, ξ and $\tilde{Q}$. In all of the simulations presented in this chapter, I fix $\beta = 1.72 \times 10^{-10}$, $\epsilon_{cor} = 10^{-12}$, $P_d = 10^{-7}$, and $N = 6.25 \times 10^{10}$.

3.7   Detector Efficiency Calibration

As mentioned above, the efficiency of the SNSPDs used in this experiment decreases as a function of increasing photon detection rates. For lower count rates (<2 Mcps), the efficiency remains close to the nominal value of 70%, but decreases with increasing count rates. This is illustrated in Fig. 3.11, where I plot the efficiency measured with a highly attenuated cw laser source (blue), as well as the pulsed source (red) used in the QKD experiment. In both cases, the efficiency decreases once the count rates exceeds 2 Mcps. The decreasing efficiency at high detection rates affect the secret key rate of the QKD system, especially at low channel loss. Although the multiplexed detection scheme reduces this effect, the secret key rate at low channel loss is still below what can be achieved with 70% efficiency single-photon detectors.

To model the rate-dependent detection efficiency in the simulation, I reorganize the data in Fig. 3.11 to plot the measured count rate as a function of expected count rate. I fit the data with a hyperbolic tangent function of the form $a \tanh(x/b) + c$, where a, b, and c are the parameters determined from the fitting, and x is the expected detection rate. Note that the parameter c should be constrained to 0 so that the model is physical. If c is non-zero, then for an expected count rate of zero, the measured count rate is positive which is not physical. The fit parameters are determined to be $a = 6.52 \times 10^6$ (6.19×10^6, 6.85×10^6) Hz and $b = 8.63 \times$

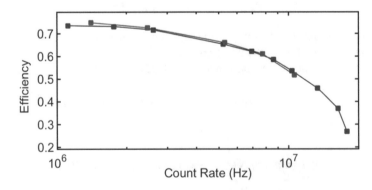

Fig. 3.11 Efficiency of single-photon detectors. Experimentally determined efficiencies of an SNSPD pixel plotted as a function of input count rate for a pulsed source (red) and a continuous-wave source (blue) at a wavelength of 1550 nm. Adapted from Islam et al. [28, e1701491, fig. S1]

10^6 (7.73×10^6, 9.53×10^6) Hz^{-1}. The detector efficiency η_d can be obtained from the fit function, which is then used to model the detector saturation across all count rates. In general, hyperbolic tangent functions are known to accurately model saturation effects of various physical processes, and for this particular set of data, I obtain a goodness of fit given by an R-squared value of 0.9963.

3.8 Numerically Optimized Secret Key Rate

To simulate the secret key presented in Fig. 3.10a, I use parameters that are experimentally determined. For example, I assume that the intrinsic error rates in time and phase basis are 3% and 2.5%, respectively, which are matched to the experimental values. I also assume that Alice transmits states in time- and phase-basis with fixed probabilities of 0.9 and 0.1, respectively.

In this section, I calculate the maximum achievable secret key rate that can be achieved with an improved experimental setup. For example, I consider that the detector efficiency is 80% and independent of photon count rates, and that the intrinsic error rates in the time and phase basis are decreased to 1% and 2%, respectively. I also assume that mean photon numbers for the signal and decoy states (μ_1 and μ_2), as well as the probability of sending time and phase states, can be tuned in the experiment. The mean photon number μ_3 is set to 0. Assuming these three quantities as free parameters, I optimize the secret key rate at every channel loss using the Matlab function `fmincon`.

Figure 3.12a shows the optimized secret key rate as a function of channel loss when Alice transmits $N = 6.25 \times 10^9$ and $N = 6.25 \times 10^{10}$ total quantum states, corresponding to a communication duration of 10 and 100 s, respectively, at a state preparation rate of 625 MHz. Figure 3.12a demonstrates that a non-zero secret key rate can be achieved at a longer distance for larger N. When $N = 6.25 \times 10^{10}$(6.25×10^9), a positive secret key rate can be obtained up to a total channel loss of 37 dB (27 dB). The optimized values of μ_1, μ_2, and p_T for $N = 6.25 \times 10^{10}$ are plotted in Fig. 3.12b. It is seen that the mean photon numbers remain approximately constant across a large channel loss, but the optimum probability of transmitting time (phase) basis decreases (increases) for larger channel losses.

3.9 Experimental Parameters

Based on the optimized probability of transmitting time basis states, in the experiment, I choose to transmit time and phase states with biased probabilities of 90% and 10%, respectively. The mean photon numbers μ_1 and μ_2 are set to 0.66 ± 0.03 and 0.16 ± 0.01, respectively, for all channel losses, except for 4 dB channel loss. For this channel loss, I set the mean photon numbers, $\mu_1 = 0.45 \pm 0.02$ and

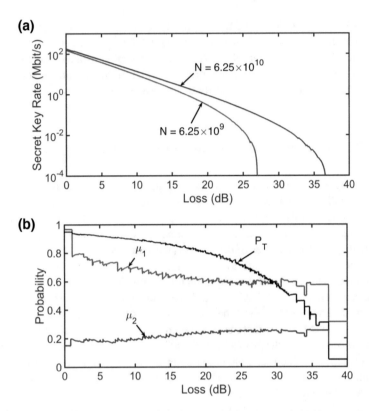

Fig. 3.12 Numerical simulation. (**a**) Optimized secret key rate as a function of loss corresponding to $N = 6.25 \times 10^9$ (red) and $N = 6.25 \times 10^{10}$ (blue). (**b**) The optimized parameters p_T, μ_1, and μ_2 for $N = 6.25 \times 10^{10}$ plotted as a function of quantum channel loss. Adapted from Islam et al. [28, e1701491, fig. S2]

$\mu_2 = 0.12 \pm 0.01$. The lower mean photon numbers are set to reduce an artifact from the readout electronics of the single-photon detector, which occurs only at high detection rates (low channel loss). Specifically, the detector pulses have some ringing due to the readout electronics. If an event is detected before the ringing of the first pulse decays, then the amplitude of the second pulse is slightly lowered, which causes a slight error in the timing of the second event. These spurious events only occur at high counts rates, and therefore I lower the mean photon numbers for the channel loss of 4 dB. At low count rates, the detection events are typically registered only after the ringing of the detector pulse decays. Therefore, the artifact is not observed when the count rate is low. Additionally, because SNSPDs operate in the Geiger mode, the amplitudes of the detector signals do not vary from one detection event to the next. Therefore, the threshold of the time-tagging module can be set to a fixed value.

Table 3.2 Sifted events and error rates registered during the 100 s of data collection

Loss (dB)	4	8	10	14	16.6
n_{T,μ_1}	3.13×10^9	1.98×10^9	1.21×10^9	5.63×10^8	2.49×10^8
n_{T,μ_2}	1.22×10^8	7.61×10^7	4.66×10^7	2.16×10^7	9.54×10^6
n_{T,μ_3}	1.58×10^7	7.15×10^6	4.46×10^6	1.67×10^6	5.83×10^5
n_{F,μ_1}	7.26×10^6	5.48×10^6	3.66×10^6	1.63×10^6	1.05×10^6
n_{F,μ_2}	4.97×10^5	3.69×10^5	2.45×10^5	1.10×10^5	4.58×10^4
n_{F,μ_3}	1.44×10^4	8.17×10^3	4.75×10^3	1.62×10^3	6.61×10^2
e_{T_1}	0.0447	0.0383	0.0373	0.0355	0.0336
e_{T_2}	0.0664	0.0521	0.0510	0.0450	0.0430
e_{F_1}	0.0478	0.0435	0.0402	0.0369	0.0485
e_{F_2}	0.0607	0.0498	0.0446	0.0488	0.0555

Table 3.3 Fractional uncertainties (%) in the measurement of the sifted data and the error rates

Loss (dB)	4	8	10	14	16.6
$\Delta n_{T,\mu_1}$	3.42	2.36	2.33	2.21	1.56
$\Delta n_{T,\mu_2}$	3.33	2.64	2.47	2.83	2.13
$\Delta n_{T,\mu_3}$	3.84	2.41	1.23	2.38	9.33
$\Delta n_{F,\mu_1}$	4.01	1.73	1.45	1.39	2.64
$\Delta n_{F,\mu_2}$	3.85	1.75	1.17	1.32	3.24
$\Delta n_{F,\mu_3}$	5.33	4.01	6.86	7.13	12.9
Δe_{T_1}	0.823	6.21	6.03	12.2	10.7
Δe_{T_2}	0.987	8.98	5.98	12.6	14.5
Δe_{F_1}	2.54	2.23	2.02	1.27	2.88
Δe_{F_2}	5.27	2.20	3.88	12.7	12.9

To minimize the effect due to the spurious events at high count rates, I set the threshold for detection to $\sim$90% of the pulse amplitude, which decreases the efficiency slightly, but also reduces the error rates. At low count rates, the threshold is set to $\sim$50%, since the artifact is not observed when the count rate is low. Finally, the mean photon number μ_3 is set to 0.002 for all losses, which is mainly due to the leakage of light through the intensity modulators.

The experimentally determined sifted counts and the corresponding error rates for all channel losses observed during 100 s of data collection are presented in Table 3.2, where I denote the error rates in time (phase) basis corresponding to mean photon number μ_k as e_{T_k} (e_{F_k}). The fractional uncertainties (%) in the measurement of the sifted data and error rates are given in Table 3.3.

3.10 Conclusion

In this chapter I demonstrate an end-to-end $d = 4$ time-phase QKD system that is capable of achieving very high secret key rates. The QKD system can obtain such high secret key rates due to multiple factors. First, for low-loss channels, the secret key throughput is constrained by detector saturation, which I overcome by using a multiplexed detection scheme and high-dimensional encoding, both of which enhance the secret key rate. Second, the system is implemented using SNSPDs, which have low timing jitter (50 ps), high quantum efficiency (nominally 70%), and high-speed readout circuits. Third, the time-bin width is set to 400 ps, which allows the system to operate at a high system clock rate of 2.5 GHz.

Other high-dimensional QKD protocols should also be able to achieve similar high rates with optimized system parameters, such as protocols based on spatial modes [14, 35], or ones that combine many photon degrees-of-freedom [15, 16]. However, this time-phase-state protocol is particularly well suited for field deployment because optical turbulence in free-space channels does not cause scattering of one of the photonic states into another as long as the wavepacket duration is substantially longer than 10 ps for path lengths of 10s of kilometers [36]. Also, in a fiber-based system, the typical dephasing time is substantially longer than the frame duration time τd.

There are several possible directions for enhancing the secret key rate of this system. One option is to use a simplified detection scheme that does not require an interferometric setup, and the complexity of the setup does not increase with the dimension of the system. Such a simplified and scalable scheme is discussed in Chap. 5. Another option is to use a monolithic on-chip interferometric receiver where many interferometers can be placed in one chip, and therefore the dimension can be increased [7, 18]. Finally, there is a significant effort underway to develop 100s of SNSPDs on one detector array, which will increase the detection rate of these systems substantially [37].

References

1. C.H. Bennett, G. Brassard, ACM Sigact News **20**, 78 (1989)
2. E. Diamanti, H.-K. Lo, B. Qi, Z. Yuan, npj Quantum Inf. **2**, 16025 EP (2016)
3. H. Bechmann-Pasquinucci, W. Tittel, Phys. Rev. A **61**, 062308 (2000). http://dx.doi.org/10.1103/PhysRevA.61.062308
4. N.J. Cerf, M. Bourennane, A. Karlsson, N. Gisin, Phys. Rev. Lett. **88**, 127902 (2002). http://dx.doi.org/10.1103/PhysRevLett.88.127902
5. L. Sheridan, V. Scarani, Phys. Rev. A **82**, 030301 (2010). http://dx.doi.org/10.1103/PhysRevA.82.030301
6. V. Scarani, H. Bechmann-Pasquinucci, N.J. Cerf, M. Dušek, N. Lütkenhaus, M. Peev, Rev. Mod. Phys. **81**, 1301 (2009). http://dx.doi.org/10.1103/RevModPhys.81.1301
7. T. Brougham, S.M. Barnett, K.T. McCusker, P.G. Kwiat, D.J. Gauthier, J. Phys. B **46**, 104010 (2013). http://stacks.iop.org/0953-4075/46/i=10/a=104010
8. J. Nunn, L.J. Wright, C. Söller, L. Zhang, I.A. Walmsley, B.J. Smith, Opt. Express **21**, 15959 (2013). http://dx.doi.org/10.1364/OE.21.015959
9. J. Mower, Z. Zhang, P. Desjardins, C. Lee, J.H. Shapiro, D. Englund, Phys. Rev. A **87**, 062322 (2013). http://dx.doi.org/10.1103/PhysRevA.87.062322
10. D.J. Gauthier, C.F. Wildfeuer, H. Guilbert, M. Stipcevic, B.G. Christensen, D. Kumor, P. Kwiat, K.T. McCusker, T. Brougham, S. Barnett, in *The Rochester Conferences on Coherence and Quantum Optics and the Quantum Information and Measurement Meeting* (Optical Society of America, Rochester, 2013), p. W2A.2. http://dx.doi.org/10.1364/QIM.2013.W2A.2

11. Z. Zhang, J. Mower, D. Englund, F.N.C. Wong, J.H. Shapiro, Phys. Rev. Lett. **112**, 120506 (2014). http://dx.doi.org/10.1103/PhysRevLett.112.120506
12. T. Brougham, C.F. Wildfeuer, S.M. Barnett, D.J. Gauthier, Eur. Phys. J. D **70**, 214 (2016). http://dx.doi.org/10.1140/epjd/e2016-70357-4
13. M. Mirhosseini, O.S. Magaa-Loaiza, M.N. OSullivan, B. Rodenburg, M. Malik, M.P.J. Lavery, M.J. Padgett, D.J. Gauthier, R.W. Boyd, New J. Phys. **17**, 033033 (2015). http://stacks.iop.org/1367-2630/17/i=3/a=033033
14. S. Etcheverry, G. Cañas, E. Gómez, W. Nogueira, C. Saavedra, G. Xavier, G. Lima, Sci. Rep. **3**, 2316 (2013)
15. G. Cañas, N. Vera, J. Cariñe, P. González, J. Cardenas, P.W.R. Connolly, A. Przysiezna, E.S. Gómez, M. Figueroa, G. Vallone, P. Villoresi, T.F. da Silva, G.B. Xavier, G. Lima, Phys. Rev. A **96**, 022317 (2017). http://dx.doi.org/10.1103/PhysRevA.96.022317
16. Y. Ding, D. Bacco, K. Dalgaard, X. Cai, X. Zhou, K. Rottwitt, L.K. Oxenlwe, npj Quantum Inf. **3**, 25 (2017)
17. J.T. Barreiro, N.K. Langford, N.A. Peters, P.G. Kwiat, Phys. Rev. Lett. **95**, 260501 (2005). http://dx.doi.org/10.1103/PhysRevLett.95.260501
18. C. Lee, D. Bunandar, Z. Zhang, G.R. Steinbrecher, P.B. Dixon, F.N.C. Wong, J.H. Shapiro, S.A. Hamilton, D. Englund, High-rate field demonstration of large-alphabet quantum key distribution (2016). http://arxiv.org/abs/arXiv:1611.01139 arXiv:1611.01139
19. J.M. Lukens, N.T. Islam, C.C.W. Lim, D.J. Gauthier, Appl. Phys. Lett. **112**, 111102 (2018). https://doi.org/10.1063/1.5024318
20. G. Ribordy, J.D. Gautier, N. Gisin, O. Guinnard, H. Zbinden, Electron. Lett. **34**, 2116 (1998)
21. C. Zhou, G. Wu, X. Chen, H. Zeng, Appl. Phys. Lett. **83**, 1692 (2003). https://doi.org/10.1063/1.1606874
22. N. Gisin, S. Fasel, B. Kraus, H. Zbinden, G. Ribordy, Phys. Rev. A **73**, 022320 (2006). http://dx.doi.org/10.1103/PhysRevA.73.022320
23. D. Hillerkuss, R. Schmogrow, T. Schellinger, M. Jordan, M. Winter, G. Huber, T. Vallaitis, R. Bonk, P. Kleinow, F. Frey et al., Nat. Photonics **5**, 364 (2011)
24. D. Hillerkuss, T. Schellinger, R. Schmogrow, M. Winter, T. Vallaitis, R. Bonk, A. Marculescu, J. Li, M. Dreschmann, J. Meyer, S.B. Ezra, N. Narkiss, B. Nebendahl, F. Parmigiani, P. Petropoulos, B. Resan, K. Weingarten, T. Ellermeyer, J. Lutz, M. Möller, M. Hübner, J. Becker, C. Koos, W. Freude, J. Leuthold, in *Optical Fiber Communication Conference* (Optical Society of America, San Diego, 2010), p. PDPC1. http://dx.doi.org/10.1364/OFC.2010.PDPC1
25. G. Thuillier, G.G. Shepherd, Appl. Opt. **24**, 1599 (1985). http://dx.doi.org/10.1364/AO.24.001599
26. Y. Hsieh, Michelson interferometer based delay line interferometers (2009), uS Patent 7,522,343. http://www.google.com/patents/US7522343
27. W.A. Gault, S.F. Johnston, D.J.W. Kendall, Appl. Opt. **24**, 1604 (1985). http://dx.doi.org/10.1364/AO.24.001604
28. N.T. Islam, C.C.W. Lim, C. Cahall, J. Kim, D.J. Gauthier, Sci. Adv. **3** (2017).http://dx.doi.org/10.1126/sciadv.1701491. http://advances.sciencemag.org/content/3/11/e1701491.full.pdf.
29. N.T. Islam, C. Cahall, A. Aragoneses, A. Lezama, J. Kim, D.J. Gauthier, Phys. Rev. Appl. **7**, 044010 (2017). http://dx.doi.org/10.1103/PhysRevApplied.7.044010
30. C. Cahall, D.J. Gauthier, J. Kim, Low-cost, low-power cryogenic readout circuit for a superconducting nanowire single photon detector system (2017). arXiv:1710.04684
31. M. Tomamichel, C.C.W. Lim, N. Gisin, R. Renner, Nat. Commun. **3**, 634 (2012)
32. M. Tomamichel, R. Renner, Phys. Rev. Lett. **106**, 110506 (2011). http://dx.doi.org/10.1103/PhysRevLett.106.110506
33. M. Lucamarini, K.A. Patel, J.F. Dynes, B. Fröhlich, A.W. Sharpe, A.R. Dixon, Z.L. Yuan, R.V. Penty, A.J. Shields, Opt. Express **21**, 24550 (2013). http://dx.doi.org/10.1364/OE.21.024550
34. T. Zhong, H. Zhou, R.D. Horansky, C. Lee, V.B. Verma, A.E. Lita, A. Restelli, J.C. Bienfang, R.P. Mirin, T. Gerrits, S.W. Nam, F. Marsili, M.D. Shaw, Z. Zhang, L. Wang, D. Englund, G.W. Wornell, J.H. Shapiro, F.N.C. Wong, New J. Phys. **17**, 022002 (2015). http://stacks.iop.org/1367-2630/17/i=2/a=022002

35. K. Brádler, M. Mirhosseini, R. Fickler, A. Broadbent, R. Boyd, New J. Phys. **18**, 073030 (2016)
36. L. Kral, I. Prochazka, K. Hamal, Opt. Lett. **30**, 1767 (2005). http://dx.doi.org/10.1364/OL.30.001767
37. M. Shaw, F. Marsili, A. Beyer, J. Stern, G. Resta, P. Ravindran, S.W. Chang, J. Bardin, F. Patawaran, V. Verma, R.P. Mirin, S.W. Nam, W. Farr, in *CLEO: 2015* (Optical Society of America, San Jose, 2015), p. JTh2A.68. http://www.osapublishing.org/abstract.cfm?URI=CLEO_SI-2015-JTh2A.68

Chapter 4
Unstructured High-Dimensional Time-Phase QKD

Conventional QKD systems require generation and detection of quantum states in two different bases, one of which is used to generate a secret key and the other is used to monitor the presence of an eavesdropper. The quantum states can be prepared in any d-dimensional Hilbert space, which means a two-basis QKD system requires generation and detection of d quantum states in the monitoring basis. Recently, Tamaki et al. [1] showed that, for a qubit-based ($d = 2$) QKD system, the protocol can be secured by transmitting two states in the information basis and only one state in the monitoring basis, while maintaining the same error tolerance against a general coherent attack as a complete setup. Here, I extend this result beyond $d = 2$ to a generic family of d-dimensional QKD protocols and show that such a system can be secured by transmitting only one state in the monitoring basis. As examples, I apply these findings to investigate the $d = 4$ time-phase QKD system demonstrated in Chap. 3 and to a $d = 7$ QKD system realized using the spatial modes of a photon [2], illustrating the applicability of this technique to various QKD schemes [3].

The numerics-based security analysis presented here is performed in collaboration with Dr. Charles Lim. Specifically, Dr. Lim shared the initial set of Matlab codes for $d = 2$, which I extend in this chapter for higher dimensional protocols and analyze the security when only a subset of monitoring basis states are transmitted. Additionally, the analysis of the spatial mode based QKD is performed with the data received from Dr. Mohammed Mirhosseini who performed the experiment demonstrated in Ref. [2].[1]

---

[1]The main results of this chapter can be found in Ref. [3].

© Springer Nature Switzerland AG 2018
N. T. Islam, *High-Rate, High-Dimensional Quantum Key Distribution Systems*,
Springer Theses, https://doi.org/10.1007/978-3-319-98929-7_4

4.1 Introduction

A conventional QKD system features a transmission device that Alice uses to generate quantum states and a receiver that Bob uses to detect the incoming quantum states. A typical two-basis qubit-based protocol requires Alice to generate two states in each of the two non-orthogonal bases, one of which is used to encode the secret key (information basis) and the other to monitor the presence of an eavesdropper (monitoring basis). For example, the information basis may consist of states $|0\rangle$ and $|1\rangle$, and the non-orthogonal monitoring basis may consist of states $|+\rangle = (|0\rangle + |1\rangle)/\sqrt{2}$ and $|-\rangle = (|0\rangle - |1\rangle)/\sqrt{2}$. The information and monitoring bases are mutually unbiased with respect to each other, which means when a state is prepared in one basis and is measured in the other, the measurement results in a completely random outcome. In the most general terms, it is the mutually unbiased property of the information and the monitoring bases that secures a QKD system against an eavesdropper.

Since the inception of the first qubit-based QKD protocol in 1984, also known as the BB84 protocol, it has been generally believed that the security of such schemes requires transmission and generation of all four quantum states in both bases. Although some theoretical investigations explored the possibility of securing a qubit-based protocol by transmitting two states in information basis and just one in monitoring basis [4], it is only recently that Tamaki et al. presented a complete security proof [1]. They showed that the security of a qubit-based protocol can be guaranteed with just three quantum states. In particular, the information basis consists of two states, $|0\rangle$ and $|1\rangle$, and the presence of Eve is determined by transmitting just one mutually unbiased basis state $|+\rangle$. They demonstrated theoretically that this simplified protocol is as robust against noise in the quantum channel as a full setup where all four quantum states are generated. Subsequently, based on their security analysis, several experimental works have demonstrated the usefulness of this protocol with simplified transmitters and receivers [5–7], thereby making the three-state protocol more attractive for practical implementations.

Although qubit-based protocols implemented with just one monitoring basis state greatly simplify QKD setups, these systems have the limitation that they can encode a maximum of one bit of information per received photon. This limitation becomes particularly important at relatively low-loss quantum channel ($<50\,\text{km}$ distances in standard optical fiber) and in QKD systems where the rate of photon generation can be significantly higher than the rate of photon detection. Indeed, many state-of-the-art transmitters allow generation of quantum states exceeding a rate of $10\,\text{GHz}$, while the maximum rate at which single photons can be detected remains below $100\,\text{MHz}$, mainly due to the long recovery time (dead time) of single-photon detectors [8].

When the rate of incoming photon states is greater than the rate at which single-photon detectors can be operated (detector saturation), many information-encoded photons are not registered by the single-photon detectors, resulting in a low secret key rate. Additionally, operating a QKD system in the detector saturation regime may lead to security loopholes (side-channel attacks) that an eavesdropper may exploit to implement attacks, such as the time-shift attack [9]. In such scenarios, it is advantageous to encode information in high-dimensional ($d > 2$) quantum states because each received photon in high-dimensional state space can encode $\log_2 d$ bits.

Recently, high-dimensional QKD systems have been widely implemented using various degrees-of-freedom, such as time-phase [10–12], orbital angular momentum (OAM) [2, 13], among others, all demonstrating significant enhancements in system performance and robustness compared to the qubit-based systems. In fact, all current state-of-the-art QKD systems capable of generating a secret key at rates exceeding 20 Mbit/s are realized using high-dimensional protocols [14, 15]. Nonetheless, the increased system performance of these protocols comes at the cost of increased complexity of the experimental setup, which makes some of these high-dimensional systems challenging to implement, especially for field applications.

Consider the specific case of the high-dimensional protocol described in Chap. 3 that uses time-phase conjugate encoding. The time states are sharply peaked single-photon wavepacket of width Δt, placed in a time bin window of width τ; a state consists of d contiguous time bins. The phase states are mutually unbiased discrete Fourier transformed states that are superposition of all the time states with distinct phases. As discussed in Chap. 3, the generation of these specific phases requires significant experimental resources, such as independent arbitrary signals, quantum random number generators, and phase modulators, which make these systems challenging to implement. Similarly, high-dimensional QKD systems implemented using OAM states require quantum random number generators, binary holograms for each state in both angular and OAM bases, and complicated detection setup to sort and detect the OAM states involving several spatial light modulators [2].

One solution to alleviate the experimental challenges is to use an unstructured version of the protocol, such as the one proposed by Tamaki et al., where only a subset of the monitoring basis states are transmitted to secure the protocol. Such a simplification retains the higher information content of high-dimensional quantum states, while simplifying the transmitter and receiver designs. Thus far, no studies in the literature have promoted the results of Tamaki et al. to higher dimensional protocols because their analytic approach is limited to one specific case where Alice transmits $d - 1$ MUB states. To prove the security of the general case where Alice can transmit any number of MUB states, it is imperative to develop new approaches to security proofs.

Recently, one such approach based on numerical methods has been used to demonstrate the security of many QKD schemes, including high-dimensional QKD protocols [16, 17]. The general theme of these analyses is to maximize Eve's mutual information with respect to Alice and Bob's quantum states using numerical optimization method. The underlying assumption in these proofs is that the statistics of Alice and Bob's compatible operators are known based on which the shared information between Alice/Bob with Eve can be maximized. Although these approaches have successfully validated the security of many existing QKD protocols, they were not used to study the case where Alice sends a subset of MUB states to secure the system.

Here, I use a similar numerical approach to promote Tamaki et al.'s result for $d = 2$ to an arbitrary d and show that the result can be generalized for any subset of the MUB states. I consider a family of two-basis high-dimensional QKD systems, where one basis is the discrete Fourier transform and mutually unbiased with respect to the other. I first analyze the security of this generic protocol against an arbitrary collective attack for the case where Alice sends complete sets of states in both bases. I then show that a complete set of monitoring basis states is not necessary to guarantee the security of this protocol. In fact, the protocol can be secured if just one state in the monitoring basis is used to determine the presence of an eavesdropper.

4.2 Protocols and Security Framework

Consider a generic QKD protocol where Alice randomly chooses a basis, T or F, and prepares a photonic wavepacket to encode a high-dimensional alphabet. The quantum states in T-basis are used to generate the secret key and the states in F-basis are used to monitor the presence of an eavesdropper. The T-basis states can be represented as $|t_n\rangle$, where $n = 0, \ldots, d - 1$. The F-basis states $|f_n\rangle$ are superposition of the T-basis states with distinct phase values determined by discrete Fourier transformation of the information basis states,

$$|f_n\rangle = \frac{1}{\sqrt{d}} \sum_{m=0}^{d-1} \exp\left(\frac{2\pi i n m}{d}\right) |t_m\rangle. \quad n = 0, \ldots, d - 1 \tag{4.1}$$

Although I use the familiar notation of time-phase QKD protocol, the analysis holds for any QKD scheme of any dimension, where two bases are related to each other via a discrete Fourier transformation.

This is an example of a prepare-and-measure scheme, in which Alice prepares the quantum states and Bob performs the measurements. It was briefly discussed in Chap. 2 that such a scheme can be written in an equivalent entanglement-based description [18], where Alice's choice of the quantum state is determined by her measurement outcome. Here, I use the entanglement-based description of the protocol to set up the optimization problem. It is important to emphasize that although I use an entanglement-based approach, both the approaches are completely equivalent.

4.2.1 The Security Framework

To set up the problem, suppose Alice and Bob share an entangled state of the form

$$|\phi\rangle_{AB} = \frac{1}{\sqrt{d}} \sum_{n=0}^{d-1} |t_n\rangle_A |t_n\rangle_B. \tag{4.2}$$

A projection measurement on the entangled state by Alice determines the state received by Bob. More precisely, suppose that Alice wants to send a state $|t_m\rangle$, $m \in \{0, \ldots, d-1\}$ to Bob. She will then project her qudit using the measurement operator, $\Pi_m = |t_m\rangle\langle t_m|$. Such a measurement will ensure that Bob receives the state $|t_m\rangle$, which now makes the protocol identical to the prepare-and-measure scenario.

Similarly, when the information is encoded in the MUB, Alice and Bob share an entangled state of the form

$$|\phi\rangle_{AB} = \frac{1}{\sqrt{d}} \sum_{n=0}^{d-1} |f_n\rangle_A |f_n\rangle_B. \tag{4.3}$$

The transformation between the T and F-bases can be performed using the Fourier operator $\mathcal{H}$ as described in Appendix C.

Additionally, suppose that Eve's interaction with the shared quantum states between Alice and Bob is independent and identically distributed (i.i.d.), which means that the quantum states are mutually independent and have identical probability distributions. After such an interaction with Alice and Bob's entangled state, the quantum state shared among Alice, Bob, and Eve can be written as $|\Psi\rangle_{ABE} = \sum_j \sqrt{\lambda_j} |\phi\rangle_{AB} |j\rangle_E$. The density matrix shared among Alice, Bob, and Eve is $\rho_{ABE} = |\Psi\rangle_{ABE}\langle\Psi|$. This is the most generic description of the shared quantum states among Alice, Bob, and Eve under the assumption of i.i.d. interaction and is also known as collective attack (see Chap. 2 for a complete definition) [18].

The primary challenge of the security analysis is to find the density matrix ρ_{ABE} that maximizes Eve's information on Alice and Bob's quantum state. This is related to the d-dimensional Shannon entropy, $H(e_F^U)$, where e_F^U is the phase error rate of the system. The phase error rate is a fictitious quantity that is not directly measured in the experiment, but rather estimated based on the other statistics. It is defined as the error rate when the entangled state is in F basis, but Alice and Bob perform the measurements in the information basis. It is known that the phase error rate is related to the amount of privacy amplification ($H(e_F^U)$) required to completely decouple the mutual information shared between Alice/Bob and Eve [19].

It is important to distinguish e_F^U from the quantum bit error rates in T and F bases defined by e_T and e_F, respectively. These are error rates that occur when Alice and Bob prepare and measure the quantum states in the same basis (T or F), but detect different quantum states.

To determine the maximum phase error rate in this protocol, I cast the problem into a maximization framework, where I use the a priori known statistics of the compatible positive-operator valued measure (POVM) of Alice and Bob. POVMs are defined as a set of positive operators (Π_m, $\langle\psi|\Pi_m|\psi\rangle \geq 0$) whose sum is the identity ($\sum_m \Pi_m = I$). For example, projection operators are a class of POVMs.

The POVM statistics can also be extracted from the experiment. For example, all the statistics of Alice and Bob's projective measurements, $\Pi_n^T = |t_n\rangle\langle t_n|$ and $\Pi_n^F = |f_n\rangle\langle f_n|$ where $n = 0$ to $d-1$, on the entangled system ρ_{AB} are well defined and can be obtained from the experiment. In addition, the statistics of the error operators E_F and E_T in T and F basis, respectively, are also known. Therefore, the problem can be expressed into the following optimization framework

$$\texttt{maximize:}\ \mathsf{Tr}(E_F \rho_{AB}) = e_F^U$$

$$\texttt{s.t.},\ \mathsf{Tr}(\rho_{AB}) = 1,$$

$$\rho_{AB} \geq 0,$$

$$\mathsf{Tr}(E_T \rho_{AB}) = e_T, \tag{4.4}$$

$$\mathsf{Tr}(\Pi_n^a \otimes \Pi_m^b \rho_{AB}) = p_{n,m}^{a,b},$$

$$\forall \{a, b\} \in \{\mathsf{T}, \mathsf{F}\}. \quad \text{and} \quad n, m = 0, \ldots, d-1$$

In Eq. (4.4), the quantum bit error rate e_T is the error rate that is measured directly in the experiment. The quantum bit error rates and the probabilities of Alice sending a state and Bob receiving a state $p_{n,m}^{a,b}$ are a priori known. The fact that I am allowing ρ_{AB} to be arbitrary also implies that Eve can perform any arbitrary operations on the states transmitted between Alice and Bob, and hence the bound is valid for collective attacks. The only unknown quantity in the optimization problem above is the phase error rate. I provide the explicit calculation for $d = 4$, including a discussion on how to determine the error operators in Appendix C. In the remaining of this section, I only discuss the important results.

4.2.2   Validation of Previously Known Results

To verify this numerical approach, I first demonstrate that this method validates the previously known results in the field. Specifically, I first calculate the secure key per transmitted state (also known as secret key fraction) as a function of the expected quantum bit error rate for the specific case of $d = 4$ under the assumption that Alice and Bob share an infinitely large ($N \to \infty$) secret key. The finite-key analysis requires additional constraints due to statistical fluctuations in the number of events received, which is omitted here for clarity. Furthermore, I also assume that Alice only transmits single-photon states. This assumption is later relaxed to incorporate the case where Alice transmits weak coherent case (see below).

Under these assumptions, the secure key fraction is given by [20]

$$K := \log_2 d - H(e_F^U) - \Delta_{\mathsf{Leak}}, \tag{4.5}$$

where $H(x) := -x \log_2(x/(d-1)) - (1-x) \log_2(1-x)$ and $\Delta_{\mathsf{Leak}} := H(e_T)$ is the fraction of key used in error correction. The secure key rate is given by $R = rK$, where r is the rate of state preparation of a d-dimensional quantum state, which may depend on d for some protocols.

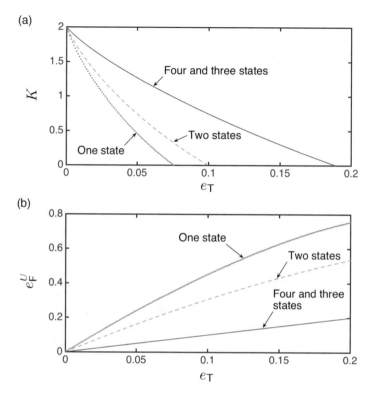

Fig. 4.1 Extractable secret key fraction with less than d monitoring states. The secret key fraction (**a**) and the numerically obtained upper bound on the phase error rate (**b**) plotted as a function of quantum bit error rate. Adapted from Ref [3], Fig. 1

In Fig. 4.1a, I plot K as a function of e_T for $d = 4$ (see Appendix C for the detailed calculation). Specifically, I consider the cases where Alice transmits four information basis states, and only a subset of monitoring basis states. For the specific case where Alice sends all four information basis states and four monitoring basis states (red line), I find that the maximum error tolerance is $\sim$18.9%, in agreement with the previously derived bounds using analytical techniques [20]. The error tolerance is defined as the maximum error rate beyond which $K = 0$. Thus, this numerical approach successfully validates the known bounds in the literature.

4.2.3   Results for Unstructured Case

The solid blue line in Fig. 4.1a, which overlaps completely with the red line, is for the case where Alice sends four information basis states, and only three monitoring basis states. This is identical to the previous case where Alice sends all four monitoring states, illustrating that one of these states is redundant. The redundancy of MUB states was previously studied analytically for qubit-type protocols in Ref. [1]. It is straightforward to find the analytical solution for this case following Ref. [1]. However, the same approach cannot be used to analyze when less than $d - 1$ monitoring states are transmitted by Alice. My numerical approach extends the result of Ref. [1], and makes it possible to analyze the security for any subset of MUB states.

For the case where Alice sends only one or two states in the monitoring basis, the protocol still generates a positive secure key fraction, as illustrated by the green and orange lines in Fig. 4.1a. However, the error rates up to which K remains positive are smaller than the four- and three-state cases. This shows that the simplification of the protocol come at the cost of reduced error tolerance.

In Fig. 4.1b, I plot the numerically obtained e_F^U as a function of e_T. As was observed for $d = 2$ [1], I find that the e_F^U for the three- and four-state cases are completely overlapping. This is because, when $d - 1$ monitoring states are transmitted, complete knowledge of the remaining unused monitoring state can be reconstructed from the statistics of $d - 1$ states that are measured, and from the statistics of events where Alice and Bob choose different bases. However, when Alice sends one or two monitoring states, complete knowledge of the non-transmitted states cannot be reconstructed, and thus the e_F^U increases faster than the e_T, resulting in a reduced secret key fraction and lower error tolerance, as shown in Fig. 4.1b.

4.2.4   Results for Three-Intensity Decoy States

The secret key fraction calculated above are for an ideal single-photon source. Most experimental (practical) implementation of QKD protocols are based on weak coherent sources that have a Poissonian photon number distribution. It is well known that a weak coherent source with multiple decoy intensities can be used to achieve secret key fraction (and rates) similar to a perfect single-photon source. I extend the above analysis to a three-intensity decoy framework in Appendix C, but the main result of that calculation is presented in Fig. 4.2.

In Fig. 4.2, I plot the secret key rate (R) as a function of the quantum channel loss for the cases where Alice transmits four, three, two, or one MUB states to secure the protocol. In this calculation, I assume that the intrinsic error rates in both the bases are 2.5%, which is close to the best experimental error rates achieved in Chap. 3. The probability of detecting a dark event in a given frame is set to 10^{-7}, which is a reasonable estimate for SNSPDs. Additionally, I assume that Alice transmits the quantum states with three different mean-photon numbers $\{\mu, \nu, \omega\}$ where $0 \leq \omega \leq \nu$ and $\nu + \omega < \mu$. To maximize the secret key rate, I optimize the mean photon numbers μ and ν at every channel loss while keeping ω constant. The optimization is carried out using the fmincon function in Matlab, where the

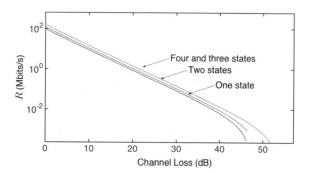

Fig. 4.2 Observation of long-distance, high-rate secret key rate with weak coherent states and decoy-state method. Achievable secret key rate as a function of channel loss for all four cases, where Alice transmits one, two, three or four states in the mutually unbiased basis to secure the protocol

mean photon numbers, μ and ν, are bounded within the range 0.01 to 0.9 photons per state under the constraint that $\nu + \omega < \mu$. From Fig. 4.2, it is observed that the secret key rate for the four- and three-state cases are perfectly overlapping, as expected from the single-photon case discussed above. When Alice transmits only one or two states, R is smaller, and the range of channel loss over which a secret key can be generated also decreases.

4.3 Applications in Various Experimental Systems

To demonstrate the applicability of this method to real QKD systems, I apply the results above to two different types of QKD systems. First, I consider the $d = 4$ time-phase QKD system discussed in Chap. 3. Second, I analyze the orbital angular momentum-based system discussed in Ref. [2].

4.3.1 Time-Phase QKD in $d = 4$

In this experiment, the time basis is the information basis and the phase basis is the monitoring basis. The time basis states are phase randomized weak coherent state wavepackets of width 66 ps, localized to one of the four contiguous time bins. The time-bin width is set to 400 ps, and hence the entire duration of the frame is 1.6 ns. The phase basis states are superposition of the time-bin states with distinct phase values as determined by the unit coefficient of the discrete Fourier transformation.

The quantum states are generated by driving an intensity and a phase modulator at a repetition rate of 625 MHz, using a sequence of arbitrary pattern of fixed length loaded on a field-programmable gate array (FPGA), as shown in Fig. 3.3. The wavepackets are then attenuated and transmitted through a quantum channel to Bob. At the receiver, Bob uses a coupler to direct 90% of the quantum states for the time-bin basis measurement and 10% for the phase basis measurement. The time basis states are measured using a low jitter single-photon-counting detector connected to a high-resolution time-to-digital converter. The detection of the phase basis states requires measuring both the time-of-arrival and the phase differences between successive peaks of the wavepacket. This is accomplished by realizing a tree-like arrangement of a cascaded time-delay interferometers, followed by single-photon-counting detectors. A detailed analysis of how a tree-like interferometric setup can be used to measure the phase states in this protocol can be found in Chap. 3.

In this experiment, all eight time and phase basis states are generated. However, the error rate in the phase basis is state-dependent, that is, some states are generated and detected more accurately than the others. This is primarily due to the experimental challenges associated with the generation of phase basis states. Specifically, to generate the states $|f_1\rangle$, $|f_2\rangle$, $|f_3\rangle$, three independent FPGA signals are amplified and combined in a 3×1 coupler. The output of the coupler is then used to drive a phase modulator which generates the distinct set of phase values required for the phase states $|f_1\rangle$, $|f_2\rangle$, $|f_3\rangle$. Although the propagation delay of each FPGA signal is matched, there is still inevitable variation in the propagation times of the signals, likely due to the variation in the cable lengths and delays in the connectors. As a result, the combined signal used to drive the phase modulator is not perfect and results in imperfect phase values. In addition, if the phase of the interferometers drifts, which can happen during a long data collection interval, it might lead to state-dependent error rates as well.

The experimentally determined error rates for each phase basis state as a function of channel loss are shown in Fig. 4.3. It is seen that the error rates for the quantum states $|f_0\rangle$ and $|f_2\rangle$ are smaller than the error rates for the remaining two states across all channel losses. Since the quantum states $|f_1\rangle$ and $|f_3\rangle$ are detected in the complementary outputs of the delay interferometer (DI 3), the increased error rates could be due to a drift in phase of this interferometer.

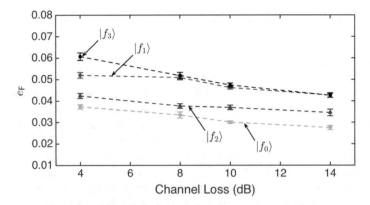

Fig. 4.3 State-dependent error rates in phase basis. The error rates corresponding to each of the phase basis states are plotted as a function of channel loss. Adapted from Ref. [3], Fig. 4(a)

Here, I consider the effect of reducing the number of phase basis states on the extractable secret key rate. Suppose, instead of sending all four phase basis states, Alice only transmits the state with the lowest error rate ($|f_0\rangle$ in this case) to determine the presence of an eavesdropper. This reduces the average quantum bit error rate in the phase basis. But, recall from Fig. 4.1b, the bound for the phase error rate when only one state is transmitted is always higher (worse) than the bound when all states are transmitted. This means that the secret key rate in the case where only one state is transmitted will be affected, even if the average quantum bit error rate is very small.

This is illustrated in Fig. 4.3, where I plot the experimentally extractable secret key rate as a function of channel loss. The orange data points represent the extractable secret key rates when all four quantum states in the phase basis are transmitted; the blue data points represent the secret key rates when only $|f_0\rangle$ is transmitted. The dashed lines represent the simulated secret key rates calculated using parameters that matches the experimental conditions.

As expected, it is seen that the secret key rate is smaller for the case where only state $|f_0\rangle$ is transmitted to monitor the presence of an eavesdropper. However, the reduction of secret key is only ~50% compared to the case where all four phase basis states are transmitted. In many implementations of QKD systems, where the highest possible secret key rate is not the most important criterion, the simplification of the experimental realization at the cost of decreased secret key rate may be beneficial (Fig. 4.4).

Specifically, consider the detailed experimental setup required to generate all eight quantum states (Fig. 4.5a) and the setup required to generate all four information basis states and just one monitoring basis state, $|f_0\rangle$ (Fig. 4.5b). To generate all states, an FPGA memory must be loaded with the random sequence of all phase values and states. Independent FPGA signals must be amplified using high-speed

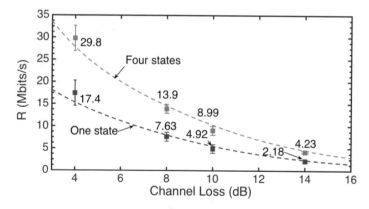

Fig. 4.4 Relatively high secret key rate with one state. Asymptotic secret key rates for the cases where Alice transmits one or four phase basis states are plotted as a function of channel loss. Adapted from Ref. [3], Fig. 4.4(b)

Fig. 4.5 Example of simplified transmitter design for the efficient QKD scheme. (**a**) The transmitter design for generating all four monitoring states is compared with (**b**) the transmitter design to generate just $|f_0\rangle$ monitoring state

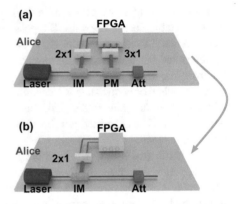

amplified (not shown for clarity) and combined using a 3×1 coupler and then drive a phase modulator. This can be quite challenging because the independent signals from the FPGA are routed along different paths, and therefore the timing offsets of each signal must be matched meticulously to ensure that the phase states are generated with high-fidelity. For the case where only $|f_0\rangle$ state is used to monitor the presence of Eve, the phase modulation part is eliminated and the setup is significantly simplified as shown in Fig. 4.5b. In fact, the same simplified setup can be used for any arbitrary d-dimensional time-phase protocol.

An important implication of this simplification is that many current QKD systems are realized with a setup similar to Fig. 4.5b, such as the coherent one-way (COW) protocol [21] and the time-bin BB84 [22]. This means that many of the existing QKD protocols can be readily reconfigured into high-dimensional schemes that can benefit from the greater information content of high-dimensional quantum states.

4.3.2   Orbital Angular Momentum-Based QKD Schemes

Here, I demonstrate the applicability of the numerics-based proof in the context of an OAM-based QKD scheme as discussed in Ref. [2].

The high-dimensional ($d = 7$) QKD system implemented in Ref. [2] is based on spatial modes of a photon. The protocol uses a photon's orbital angular momentum (OAM) modes and the azimuthal angles (ANG) as the two non-orthogonal bases. The states in angular and OAM bases are related to each other via the transformation:

$$\Psi^n_{\text{ANG}} = \frac{1}{\sqrt{d}} \sum_{\ell=-N}^{N} \Psi^\ell_{\text{OAM}} \exp\left(\frac{i 2\pi n \ell}{d}\right). \tag{4.6}$$

By manipulating the index ℓ, it is possible to show that this representation of the ANG basis states is identical to the previously discussed transformation given by Eq. (4.1). Therefore, the formalism developed in Sect. 4.2 is also applicable to this protocol.

The experimental demonstration of this protocol is performed with signals of fixed intensity (no decoy states). To maximize the probability of having one photon per transmitted state, all signals are attenuated to have a mean photon number of 0.1 per state. Based on the detection statistics, a quantum bit error rate of 10.5% is determined. Although the reported error rate is 10.5%, I find that the quantum bit error rate is 10.9% based on my analysis of the raw data provided by the authors. Approximately 4% of the error rate can be attributed to the high detector dark count rate, and the remaining 6.5% to the intermodal cross-talk. After taking into account the error correction and the privacy amplification, they calculate a secret key fraction of $\log_2(7) - 2H(0.105) = 1.29$ bits/photon, which is substantially lower than the theoretical limit of $\log_2(7) = 2.81$ bits/photon.

To analyze this spatial-modes-based QKD scheme in the context of the asymmetric protocol described here, suppose that the OAM basis is used to encode information and the ANG basis is used to monitor the presence of an eavesdropper. The average error rates in the OAM basis (e_{OAM}) and the ANG basis (e_{ANG}) are 9.4% and 12.2%, respectively. The error rate for each ANG states is shown in Table 4.1. If the protocol is realized using only a subset of the ANG basis, then the average error rate for those subset of states would be smaller.

Figure 4.6 shows the dependence of K on the error rate e_{ANG} for all seven monitoring basis states. The dashed lines correspond to the error rates that would be obtained if only a subset of the ANG basis states were used to secure the system. It is seen that when six states are transmitted to monitor the presence of Eve, the secret key fraction is higher than the case when all seven states are used, illustrating that sending one less state in this case would be useful. When less than six states are transmitted, K decreases substantially. Interestingly, when only two states are transmitted, the secret key fraction is higher than the case when five states are transmitted. This illustrates that, in addition to simplifying the experimental setup, sometimes transmitting fewer states can generate higher secret key fraction (secret key rate).

Table 4.1 Error rates in ANG basis

n of $|\Psi_{ANG}^n\rangle$	e_{ANG}
7	0.189
6	0.197
5	0.048
4	0.049
3	0.111
2	0.071
1	0.200

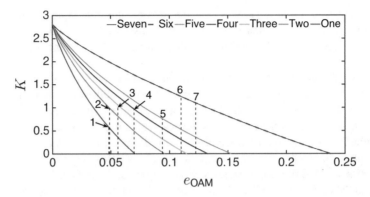

Fig. 4.6 Secret key fraction for $d = 7$ OAM-based QKD. The dependence of K on e_{ANG} when Alice transmits up to seven MUB states are shown

4.4 Conclusion

In this chapter, I describe a numerics-based security analysis that can be applied to a family of two-basis arbitrary dimensional QKD protocols. I show that such a protocol can be secured with a subset of mutually unbiased basis states to monitor the presence of an eavesdropper, which offers significant advantages in simplifying the experimental setup of high-dimensional QKD systems. Another implication of this method is that many of the current QKD systems such as the time-bin-encoding qubit-based protocols can be easily upgraded to high-dimensional QKD schemes with simple modifications to the setup, and will therefore increase the secure key rate of these systems. This work significantly simplifies the transmitter designs of many QKD systems, including those using spatial (orbital-angular momentum or polarization), time-phase, etc., modes of a photon, making these systems easier to implement with low resources and potentially at lower cost.

References

1. K. Tamaki, M. Curty, G. Kato, H.-K. Lo, K. Azuma, Phys. Rev. A **90**, 052314 (2014). http://dx.doi.org/10.1103/PhysRevA.90.052314
2. M. Mirhosseini, O.S. Magaa-Loaiza, M.N. OSullivan, B. Rodenburg, M. Malik, M.P.J. Lavery, M.J. Padgett, D.J. Gauthier, R.W. Boyd, New J. Phys. **17**, 033033 (2015). http://stacks.iop.org/1367-2630/17/i=3/a=033033
3. N.T. Islam, C.C.W. Lim, C. Cahall, J. Kim, D.J. Gauthier, Phys. Rev. A **97**, 042347 (2018). http://dx.doi.org/10.1103/PhysRevA.97.042347
4. C.-H.F. Fung, H.-K. Lo, Phys. Rev. A **74**, 042342 (2006). http://dx.doi.org/10.1103/PhysRevA.74.042342
5. F. Xu, K. Wei, S. Sajeed, S. Kaiser, S. Sun, Z. Tang, L. Qian, V. Makarov, H.-K. Lo, Phys. Rev. A **92**, 032305 (2015). http://dx.doi.org/10.1103/PhysRevA.92.032305

6. Z. Tang, K. Wei, O. Bedroya, L. Qian, H.-K. Lo, Phys. Rev. A **93**, 042308 (2016). http://dx. doi.org/10.1103/PhysRevA.93.042308
7. D. Bunandar, A. Lentine, C. Lee, H. Cai, C.M. Long, N. Boynton, N. Martinez, C. DeRose, C. Chen, M. Grein, D. Trotter, A. Starbuck, A. Pomerene, S. Hamilton, F.N.C. Wong, R. Camacho, P. Davids, J. Urayama, D. Englund, Phys. Rev. X **8**, 021009 (2018). http://dx. doi.org/10.1103/PhysRevX.8.021009
8. F. Marsili, V.B. Verma, J.A. Stern, S. Harrington, A.E. Lita, T. Gerrits, I. Vayshenker, B. Baek, M.D. Shaw, R.P. Mirinet al., Nat. Photonics **7**, 210 (2013)
9. B. Qi, C.-H.F. Fung, H.-K. Lo, X. Ma, Quantum Inf. Comput. **7**, 73 (2007)
10. T. Brougham, S.M. Barnett, K.T. McCusker, P.G. Kwiat, D.J. Gauthier, J. Phys. B **46**, 104010 (2013). http://stacks.iop.org/0953-4075/46/i=10/a=104010
11. T. Brougham, C.F. Wildfeuer, S.M. Barnett, D.J. Gauthier, Eur. Phys. J. D **70**, 214 (2016). http://dx.doi.org/10.1140/epjd/e2016-70357-4
12. N.T. Islam, C. Cahall, A. Aragoneses, A. Lezama, J. Kim, D.J. Gauthier, Phys. Rev. Appl. **7**, 044010 (2017). http://dx.doi.org/10.1103/PhysRevApplied.7.044010
13. J. Leach, E. Bolduc, D.J. Gauthier, R.W. Boyd, Phys. Rev. A **85**, 060304 (2012). http://dx.doi. org/10.1103/PhysRevA.85.060304
14. C. Lee, D. Bunandar, Z. Zhang, G.R. Steinbrecher, P.B. Dixon, F.N.C. Wong, J.H. Shapiro, S.A. Hamilton, D. Englund, High-rate field demonstration of large-alphabet quantum key distribution (2016). arXiv:1611.01139
15. N.T. Islam, C.C.W. Lim, C. Cahall, J. Kim, D.J. Gauthier, Sci. Adv. **3** (2017). http://dx.doi.org/ 10.1126/sciadv.1701491. http://advances.sciencemag.org/content/3/11/e1701491.full.pdf
16. P.J. Coles, E.M. Metodiev, N. Lütkenhaus, Nat. comm. **7**, 11712 (2016)
17. K.T. Goh, J.-D. Bancal, V. Scarani, New J. Phys. **18**, 045022 (2016)
18. V. Scarani, H. Bechmann-Pasquinucci, N.J. Cerf, M. Dušek, N. Lütkenhaus, M. Peev, Rev. Mod. Phys. **81**, 1301 (2009). http://dx.doi.org/10.1103/RevModPhys.81.1301
19. M. Koashi, New J. Phy. **11**, 045018 (2009). http://stacks.iop.org/1367-2630/11/i=4/a=045018
20. L. Sheridan, V. Scarani, Phys. Rev. A **82**, 030301 (2010). http://dx.doi.org/10.1103/PhysRevA. 82.030301
21. B. Korzh, C.C.W. Lim, R. Houlmann, N. Gisin, M.J. Li, D. Nolan, B. Sanguinetti, R. Thew, and H. Zbinden, Nat. Photonics **9**, 163 (2015). http://dx.doi.org/10.1038/nphoton.2014.327
22. M. Lucamarini, K.A. Patel, J.F. Dynes, B. Fröhlich, A.W. Sharpe, A.R. Dixon, Z.L. Yuan, R.V. Penty, A.J. Shields, Opt. Express **21**, 24550 (2013). http://dx.doi.org/10.1364/OE.21.024550

Chapter 5
Scalable High-Dimensional Time-Bin QKD

5.1 Introduction

The QKD system demonstrated in Chap. 3 is based on a four-dimensional time-phase encoding scheme in which time basis states are used to generate a secret key and phase basis states are used to monitor the presence of an eavesdropper. The system can generate a secret key at a high rate, mainly due to the greater information content of the high-dimensional quantum states and a multiplexed detection scheme. Given the success of the $d = 4$ protocol, an important subsequent question to address is: does increasing the dimension beyond $d = 4$ increase the secret key rate under realistic experimental conditions?

In general, higher dimensional protocols encode more bits of information per photon, scaling logarithmically as $\log_2 d$. When the high-dimensional quantum states are generated using time-bin encoding scheme, there is an additional $1/d$ scaling factor in the rate (bits/second) that arises due to the increased frame size. When these systems are operated at short-to-moderate distances, corresponding to relatively low quantum channel loss, the saturation of single-photon detectors remains a limiting factor and the additional $1/d$ does not affect the secret key rate of the system. Therefore, increasing the dimension of the encoding states enhances the secret key rate. The protocol is also scalable to any arbitrary dimension, although the measurement scheme required to detect d-dimensional phase states involves $2d - 1$ interferometers that are costly and require moderate temperature stabilization as shown in Appendix B.

In this chapter, I demonstrate a simplified receiver design consisting of only one beamsplitter, a local source of quantum states and two single-photon counting modules that can be used to measure phase states of any dimension. This technique relies on the interference of two identical photons, also known as the Hong-Ou-Mandel (HOM) effect, and has been used in the past to measure the phase values of quantum photonic wavepackets in several QKD demonstrations, such as the measurement-device independent [7] and the round-robin QKD [3] schemes. Here,

© Springer Nature Switzerland AG 2018
N. T. Islam, *High-Rate, High-Dimensional Quantum Key Distribution Systems*,
Springer Theses, https://doi.org/10.1007/978-3-319-98929-7_5

I show that it is possible to measure the phase values of high-dimensional quantum states using a similar technique.

This chapter is outlined as follows: for pedagogical reasons, I start with a discussion of HOM interference of two identical photons. Then, I show qualitatively that an interferometric measurement of the phase basis states is equivalent to a two-photon HOM interference measurement. I then discuss the security of the protocol and simulate the secret key rates considering non-ideal experimental parameters. Finally, I describe the experimental realization of the system involving two independent laser sources, and multi-intensity transmitted states (decoy method). I then conclude the chapter with a summary and future improvements that can potentially increase the secret key rate.

5.2   The Hong-Ou-Mandel Interference

Consider two single-photon wavepackets simultaneously incident at a 50/50 beamsplitter, each arriving from a different input port of the beamsplitter as shown in Fig. 5.1. The interference of the wavepackets as observed from the correlated events in detectors $D0$ and $D1$ depends on the distinguishability of the two incident photons.

First, consider the case where both the photons incident at the beamsplitter are perfectly distinguishable and are simultaneously incident at the beamsplitter as shown in Fig. 5.2. There are four possible outcomes that can be observed at the single-photon counting detectors $D0$ and $D1$ as shown in Fig. 5.2a–d. The photons can reflect or transmit in two different directions as illustrated in Fig. 5.2a and b, in which case a coincidence event is observed in the detectors $D0$ and $D1$. The remaining possibilities are the cases where both the photons bunch and go towards the same detector as shown in Fig. 5.2c and d. For these bunched events, no correlated events are observed in detectors $D0$ and $D1$.

Fig. 5.1 Schematic illustration of two-photon interference setup. Two photons are simultaneously incident at two input ports of a 50/50 beamsplitter and the output ports of the beamsplitters are coupled into two single-photon counting modules

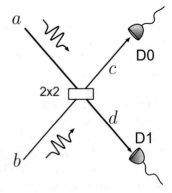

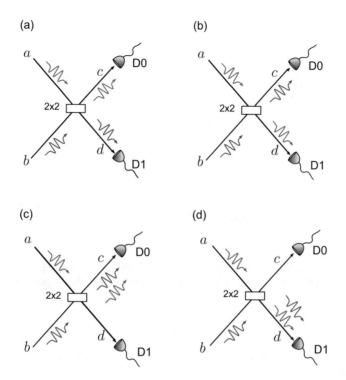

Fig. 5.2 Interference of distinguishable photons at a beamsplitter. The two photons can either be reflected (**a**) or transmitted (**b**) towards different detectors, or observed in the same detector (**c**) and (**d**)

The outcomes are surprisingly different when the incident photons are completely indistinguishable (identical): No correlated events are observed in the detectors $D0$ and $D1$, indicating that the two photons never exit from different ports of the beamsplitter as shown in Fig. 5.3a and b. This coalescence of photons, known as the Hong-Ou-Mandel (HOM) effect [4], arises from the destructive interference between the two quantum mechanical probability amplitudes of the photons. When the wavepackets do not overlap temporally, interference between the two wavepackets is not possible. Hence the possibilities are identical to the case where the two photons are perfectly distinguishable (Fig. 5.2), and correlated events are observed in the detectors about half the time.

The HOM effect can be illustrated mathematically by considering the transformation of a single-photon field through a beamsplitter. A beamsplitter transforms an input field $a^\dagger$ representing a single-photon incident at the input port a (Fig. 5.1) as

$$a^\dagger \xrightarrow{\text{BS}} \frac{1}{\sqrt{2}}(c^\dagger + d^\dagger), \tag{5.1}$$

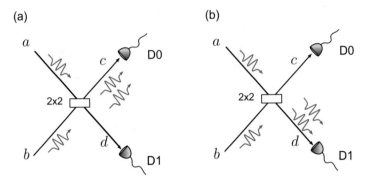

Fig. 5.3 Interference of indistinguishable photons at a beamsplitter. The photons bunch together and are recorded in the same detector, either $D0$ (**a**) or $D1$ (**b**). No correlated events are observed in detectors $D0$ and $D1$

where $c^\dagger$ and $d^\dagger$ represent the fields at the output ports of the beamsplitter. Similarly, the transformation of a single-photon field incident at the other input port b is

$$b^\dagger \xrightarrow{\text{BS}} \frac{1}{\sqrt{2}}(c^\dagger - d^\dagger). \tag{5.2}$$

Therefore, the transformation of the two input photons at the beamsplitter can be written as

$$a^\dagger b^\dagger \xrightarrow{\text{BS}} \frac{1}{2}(c^\dagger + d^\dagger)(c^\dagger - d^\dagger), \tag{5.3}$$

$$= \frac{1}{2}(c^\dagger c^\dagger - c^\dagger d^\dagger + d^\dagger c^\dagger - d^\dagger d^\dagger). \tag{5.4}$$

The bosonic nature of photons requires the commutation relation $[c^\dagger, d^\dagger] = 0$. Therefore, the two photons incident at the beamsplitter transform as

$$a^\dagger b^\dagger \xrightarrow{\text{BS}} \frac{1}{2}(c^\dagger c^\dagger - d^\dagger d^\dagger). \tag{5.5}$$

From the relation above, it is clear that when two identical single-photon states are incident simultaneously from two different input ports of a beamsplitter they coalesce. Half the time they exit from the output port c, and half the time they exit from the output port d, resulting in no correlated events in detectors $D0$ and $D1$.

When the single-photon wavepackets in a HOM interference setup are replaced with phase-randomized weak coherent-state (PRWCS) wavepackets, coincidence events are observed in the detectors. This is because the photon number of a highly attenuated coherent laser is dictated by a Poisson distribution. As a result, only a small fraction of the states generated from a weak coherent source contains exactly one photon and a large number of states contain a zero photon (vacuum) or more than one photon, which results in imperfect two-photon interference.

Nonetheless, the rate of coincidence events is only half compared to the perfectly distinguishable states as shown in Fig. 5.2, indicating that the quantum nature of light can also be observed with an imperfect source such as PRWCS. The reduction is only 50% because there are cases when a vacuum is incident from one port of the beamsplitter and two or more photons are incident from the other, leading to a 50% probability of observing a coincidence event. Additionally, there are also cases when one or more photons are incident from both input ports of the beamsplitter, resulting in correlated events at the detectors.

Although HOM interference with a PRWCS is imperfect, these states can still be used for experiments that require pure single-photon states using the familiar technique of decoy states. Some examples of this method are illustrated in Chaps. 3 and 4, where weak coherent sources of three distinct intensities are used to extract single-photon statistics. A similar technique can also be used here, except a generalization of the decoy-method is required to extract one photon events from both Alice and Bob. Over the last few years, extensive analyses of such decoy methods have been developed to analyze measurement device independent QKD protocols [5, 7–10]. More recently, Yuan et al. [11] and Navarrete et al. [6] generalized this method for an $M \times N$ beamsplitter, primarily to analyze intricate quantum experiments, such as the Boson sampling problem. Here, I use results from these studies to analyze this new phase state measurement scheme.

5.3 Phase State Measurement Scheme with a Local Oscillator

The HOM interference is a quantum phenomenon that is not observed with classical photonic wavepackets. This provides an efficient and a simple means of measuring phase states in a time-phase QKD. In this section, I explain how this technique can be used to measure phase states in a high-dimensional time-phase QKD system.

The phase state measurement device discussed in Chap. 3 requires $2d - 1$ interferometers to measure all d phase states. For a four-dimensional protocol, the tree-like arrangement consists of three time-delay interferometers with the optical path-length differences and the phases set so that there is a one-to-one mapping between the input phase state $|f_n\rangle$ and the detector D_n that registers the event (Fig. 5.4a). The overall effect of the interferometric setup is to interfere every peak of the single-photon wavepacket with a successive peak, and based on the detector in which the event is recorded, it is then possible determine if a correct phase state is detected.

An equivalent method of measuring the phase values is to interfere an incoming phase state wavepacket with a second wavepacket of identical global phase δ as shown in Fig. 5.4b. In general, wavepackets representing the phase states have two different types of phases: the local phase that are the distinct phase values determined from the Fourier transformation, and the global phase which is an arbitrary phase that is constant across the entire frame. For example, consider the state $|f_2\rangle$ in $d = 4$ that have local phase values of 0, π, 2π, 3π. If the global

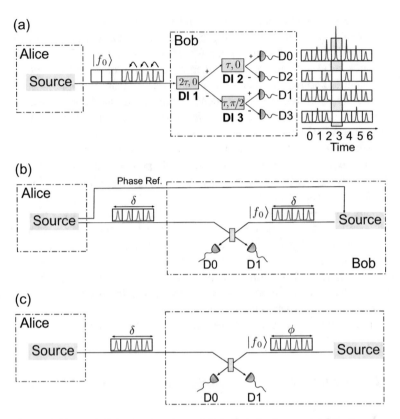

Fig. 5.4 Comparison of different phase state measurement schemes. (**a**) The interferometric method requires $2d - 1$ time-delay interferometers to measure a phase state of dimension d. (**b**) A second method is to phase lock a source at Bob's receiver with respect to Alice's source, and generate identical wavepackets and interfere the incoming quantum states with the locally generated quantum state. (**c**) Another way to measure the phase states is to generate the local quantum state "locally" with no phase locking and perform a HOM-type interference

phase of the frame is δ, then the absolute phases of the wavepacket peaks are: $\delta + 0, \delta + \pi, \delta + 2\pi, \delta + 3\pi$. If Bob has a source that generates quantum states of the same global phase as Alice's, then he could in principle just interfere the incoming quantum states from Alice with a state that he generates of the same global phase. Depending on the phase value of each peak, a constructive, destructive or no interference is observed in every time bin, which then allows a perfect discrimination of all the phase states.

As an example, consider the specific case where the incoming wavepacket from Alice is $|f_2\rangle$. If Bob generates the quantum state $|f_0\rangle$ with the same global phase as Alice's $|f_2\rangle$, and interferes the two states at a beamsplitter, then the detector D0 (D1) will record constructive (destructive) interference in alternative time bins. This pattern of probability distribution is unique for each of the d phase states that Alice transmits, and hence a perfect discrimination of the phase states is possible.

One challenge with this approach is that distribution of phase reference over a quantum channel can lead to new side-channel attacks, such as a Trojan horse attack [2]. In general, it is not necessary for the wavepackets to have identical global phase. In fact, the global phases of the interfering wavepackets can be completely random in which case the interference will be HOM-type. For such a measurement scheme, the incoming wavepacket of global phase δ can be interfered with a locally generated wavepacket of global phase ϕ in Bob's setup (Fig. 5.4c). A successful measurement of the phase state is determined from the clicks registered in detectors $D0$ and $D1$. If the two photons are identical, no correlated events are observed in detectors $D0$ and $D1$. If any disturbance in the quantum channel perturbs the relative local phases between the wavepacket peaks, the two states are no longer identical and a coincidence event is recorded in the two detectors.

The interference of the incoming wavepacket with a locally generated quantum state is an active way of determining whether the incoming wavepacket from Alice matches with Bob's choice of the measurement state. I refer to this scheme as a quantum-controlled QKD protocol because the measurement of quantum states is performed with an ancilla (local) quantum state generated in Bob's measurement setup.

It is important to realize that the local oscillator and the interferometric-based measurement schemes are equivalent techniques for measuring phase states in a QKD protocol (for a more rigorous proof see Ref. [3]). The protocol is also carried out in a similar way: Alice generates quantum states in a randomly chosen basis and transmits the states to Bob via a quantum channel; Bob uses a beamsplitter to measure a fraction of the incoming states in the time basis and the remaining in the phase basis. For a time basis measurement, Bob measures the time-of-arrival of the incoming photons. For a phase basis measurement, he generates an identical phase state of randomized global phase and overlaps the incoming state with the locally generated state, and records the pattern of detector clicks. An event is considered to be an error if Alice and Bob both send the same phase state and Bob's detectors record a coincidence. In the end of the communication session, Alice and Bob communicate over a public channel and discuss their basis choice. They then perform the post-processing steps of sifting, error verification and correction, and privacy amplification

5.3.1 Efficiency Comparison with the Interferometric Approach

As discussed in Chaps. 2 and 3, when a d-dimensional phase state is measured using an interferometric setup, the probability that the network of interferometers will conclusively detect the phase state is $1/d$. The large inefficiency of the detection mechanism arises due to the spilling of the wavepacket to adjacent time bins, which leads to inconclusive events (see Appendix A).

The efficiency of the quantum-controlled approach depends on the number of phase basis states that are used to secure the QKD scheme. In the case where all d states are generated, Bob has to generate all d phase states as well. As a result, the probability that Bob will choose the same state as Alice is only $1/d$, which is identical to the interferometric setup. Therefore, the efficiency of the local oscillator-based detection scheme is identical to the interferometric-based scheme if all the phase states are used to monitor the presence of an eavesdropper.

Another possibility is to implement this protocol using the efficient QKD scheme described in Chap. 4, where Alice sends all d time basis states and only a subset of the phase basis states. In the case where Alice chooses to send just one phase basis state, the efficiency of phase basis measurement at Bob's receiver is 100%, although the tolerance for quantum bit error rate decreases, and the bound for the phase error rate is worse as discussed in Chap. 4. If the QKD system has a low error rate in the phase basis states (see below), then sending just one state is the most effective way to generate a secret key.

It is important to note that the efficiency of the interferometric setup does not increase when Alice generates only a subset of the quantum states. This is because, unlike the quantum-controlled technique, the interferometric method is passive and requires the same number of time-delay interferometers, which in turn results in the same inefficiency as the complete protocol. Due to this advantage, I implement the local oscillator-based scheme with just one phase state, which also greatly simplifies the experimental setup.

5.4   Security Analysis

The security analysis of this protocol is derived from the numerics-based proof developed in Chap. 4. At the source, Alice generates the time and phase basis states as discussed in earlier chapters. At the receiver, the time basis states are measured using a single-photon counting module and a high-speed time-to-digital converter. The only difference is in the measurement of the phase basis states, but it can be shown that the 2-photon HOM interference measurement is equivalent to the interferometric measurement [3]. Therefore, the same security analysis can be applied to this protocol.

In the remaining of this section, I simulate the extractable secret key rate in the limit of an asymptotically large key length and for the case where Alice and Bob send weak coherent states of different mean photon numbers (decoy method). Since this is a relatively new protocol, the finite-key security analysis of this protocol remains an open problem. However, a good starting point for such an analysis would be to utilize the finite-key results from the MDI-QKD [1]. This will be addressed in a future work.

5.4.1 Simulated Secret Key Rate

I simulate the secret key rate of this protocol assuming that Alice sends weak coherent states of three different intensities, $k \in \{\mu, \nu, \omega\}$ in the time and phase basis with probabilities p_T and $p_F := 1 - p_T$, respectively. Each of the mean photon numbers is transmitted with probability p_k. To calculate the extractable secret key rate, it is important to bound the single-photon gain in both bases, as well as the error rate in phase basis as a function of the quantum bit error rate. The definitions of yield, gain and error rates are discussed in Appendix C. For completeness, I summarize them here.

Briefly, the i-photon gain in the time basis is defined as the joint probability of Alice transmitting an i-photon state and Bob receiving a detection event. This can be expressed as

$$R_{T,i} = Y_{T,i} \frac{k^i}{i!} \exp(-k), \tag{5.6}$$

where $Y_{T,i}$ is the yield of i-photon state in the time basis. The yield is defined as the conditional probability of Bob receiving a detection event given Alice sends an i-photon state. The overall gain can be expressed as

$$R_{T_k} = \sum_{i=0}^{\infty} Y_{T,i} \frac{k^i}{i!} \exp(-k), \tag{5.7}$$

which is the joint probability that Alice sends a state with mean photon number k and Bob receives a detection event. This quantity is directly measured in the experiment by calculating the total number of detection events per total signals transmitted. The goal of the decoy-state analysis is to estimate the single-photon yield ($Y_{T,1}$) and the single-photon gain ($R_{T,1}$), which are not directly accessible through experimental data. By measuring the overall gain for each of the three mean photon numbers, and manipulating the expressions above, it is possible to obtain a tight bound for the single-photon yield and gain (see Appendix C).

In the phase basis, these definitions are promoted to take into account that both Alice and Bob send quantum states. Specifically, the i, j-photon gain is defined as the probability of Alice sending an i-photon state and Bob a j-photon state, and Bob's detectors registering a coincident event. The i, j-photon gain can be expressed as

$$R_{F,ij} = Y_{F,ij} \frac{k_A^i k_B^j}{i! j!} \exp[-(k_A + k_B)], \tag{5.8}$$

where $k_A(k_B)$ is the mean photon number of Alice (Bob); $Y_{F,ij}$ is the conditional probability of receiving a coincidence detection given Alice sends an i-photon state and Bob sends a j-photon state. The overall gain of Alice's mean photon number k_A and Bob's mean photon number k_B can be written as

$$R_{F_{k_A k_B}} = \sum_{i,j=0}^{\infty} Y_{F,ij} \frac{k_A^i k_B^j}{i!j!} \exp[-(k_A + k_B)]. \tag{5.9}$$

Using these definitions as the starting point, the lower-bound secret key fraction can be written as

$$\ell/N := R_{T,1}^L [\log_2 d - H(e_F^U)] - R_T H(e_T), \tag{5.10}$$

where ℓ is the length of secret key string; N is the total number of quantum states transmitted by Alice; $R_{T,1}^L$ is the lower bound of single-photon gain in time basis; e_F^U is the upper bound on the *phase error rate* obtained when Alice sends a phase state and Bob sends a phase state; $R_T := (p_\mu R_{T_\mu} + p_\nu R_{T_\nu} + p_\omega R_{T_\omega})$ is the gain of time basis states weighting contribution from all mean photon numbers; e_T is the quantum bit error rate in time basis; $H(x)$ is the Shannon entropy for d-dimensional quantum states.

The lower bound of single-photon yield and gain in time basis was previously derived in Chap. 4 (Appendix C) and omitted here for clarity. The lower bound of two-photon yield can be written as [9]

$$Y_{F,11} \geq \frac{1}{(\mu - \omega)^2 (\nu - \omega)^2 (\mu - \omega)} \tag{5.11}$$

$$\times [(\mu^2 - \omega^2)(\mu - \omega)\{R_{F_{\nu\nu}} \exp(2\nu) + R_{F_{\omega\omega}} \exp(2\omega) - R_{F_{\nu\omega}} \exp(\nu + \omega)$$

$$- R_{F_{\omega\nu}} \exp(\omega + \nu)\}$$

$$- (\nu^2 - \omega^2)(\nu - \omega)\{R_{F_{\mu\mu}} \exp(2\mu) + R_{F_{\omega\omega}} \exp(2\omega) - R_{F_{\mu\omega}} \exp(\mu + \omega)$$

$$- R_{F_{\omega\mu}} \exp(\omega + \mu)\}],$$

where $R_{F_{k_A k_B}}, k_A, k_B \in \{\mu, \nu, \omega\}$ is the gain when Alice sends a state of mean photon number k_A and Bob sends k_B. Similarly, the upper bound for the single-photon error rate is given by

$$e_{F,11} \leq \frac{1}{(\nu - \omega)^2 Y_{F,11}} \times [\exp(2\nu) R_{F_{\nu\nu}} e_{\nu\nu} + \exp(2\omega) R_{F_{\omega\omega}} e_{\omega\omega} \tag{5.12}$$

$$- \exp(\nu + \omega) R_{F_{\nu\omega}} e_{\nu\omega} - \exp(\omega + \nu) R_{F_{\omega\nu}} e_{\nu\omega}],$$

where $e_{k_A k_B}$ is the *observed* quantum bit error rate when Alice sends a state with a mean photon number of k_A and Bob sends a state with a mean photon number with k_B. Note that, when Alice and Bob transmit all d or $d - 1$ states, $e_F^U = e_{F,11}$ as discussed in Chap. 4. Otherwise, the e_F^U can be determined by first calculating $e_{F,11}$, and then using this bound to find the corresponding e_F^U from the numerical optimization as discussed in Chap. 4. Incorporating all the theoretical tools, it is now possible to simulate the secret key rate in the asymptotic limit.

To promote this result into finite-key regime, one has to take into account the statistical deviation of each gain term. The statistical deviation is typically calculated from the Hoeffding-Chernoff inequality as was done in Chap. 3 for the $d = 4$ time-phase QKD protocol. The end result of including these terms is that each of the final bounds of $Y_{F,11}$ and e_F^U will have one additional term that will take the finite-key statistical fluctuations into account and make the bounds tighter. Although some recent investigations of MDI-QKD schemes have used the finite-key results for security analyses [1, 10], the extension of Eqs. (5.11) and (5.12) to finite-key limit is still a challenge. This is primarily because of the hybrid nature of this protocol where measurement in one basis involves a two-photon interference similar to an MDI-QKD experiment, and the measurement in the other basis is similar to a prepare-and-measure scheme.

One important caveat for simulating secret key rates is to properly estimate the error rates $e_{k_A k_B}$. For $k_A = k_B$, the intrinsic error rate is 0.25. This is due to the imperfect HOM interference of weak coherent states. More precisely, a valid detection is obtained when there are at least two photons that are detected at the output(s) of the beamsplitter. For small mean photon numbers, this may happen due to three specific cases: (1) one photon comes from Alice and one from Bob; (2) two photons come from Alice and zero from Bob; (3) two photons come from Bob and zero from Alice.

When two photons—one from Alice and one from Bob—arrive at the beamsplitter, the theoretical error rate is zero, assuming that the photons are perfectly indistinguishable. When two photons come from Alice or Bob and zero from the other, there is a 50% chance that the photons will result in a coincidence at the detectors and therefore cause an error. Suppose Alice transmits her quantum states with mean photon number k_A and Bob with mean photon number k_B. Then the probability of obtaining two photons from Alice or Bob, and zero from the other can be written as

$$P_{2,0} + P_{0,2} = \frac{\exp(-k_A - k_B)k_A^2}{2!} + \frac{\exp(-k_B - k_A)k_B^2}{2!} \tag{5.13}$$

Likewise, the probability of obtaining a photon from Alice and one from Bob can be written as

$$P_{1,1} = \exp(-k_A)k_A \exp(-k_B)k_B. \tag{5.14}$$

The mean error rate is given by

$$e_{k_A k_B} := \frac{(P_{2,0} + P_{0,2})/2 + 0(P_{1,1})}{P_{1,1} + P_{0,2} + P_{2,0}}. \tag{5.15}$$

For the specific case where $k_A = k_B = \mu$,

$$e_{\mu\mu} := \frac{[\exp(-2\mu)\mu^2]/2}{\exp(-2\mu)\mu^2 + \exp(-\mu)\mu^2}, \tag{5.16}$$

which in the limit when $\mu \ll 1$ approaches 0.25. The error rate of 0.25 is also obtained for e_{vv} and $e_{\omega\omega}$. This is the minimum error rate that can be achieved with PRWCS. When $k_A \neq k_B$, the intrinsic error rate can be determined by substituting the specific mean photon numbers in Eqs. (5.13) and (5.14) and then calculating the error rate from Eq. (5.15).

In an experiment, it is difficult to measure the quantities $P_{2,0}$, $P_{0,2}$, and $P_{1,1}$. So, the error rates need to be related to quantities that can be measured in a laboratory. To this end, I define the HOM visibility as

$$\mathcal{V}_{\text{HOM}} := \frac{P_C^d - P_C}{P_C^d} = 1 - \frac{P_C}{P_C^d}, \tag{5.17}$$

where P_C^d is the probability of seeing a coincidence when the wavepackets are completely distinguishable and P_C is the probability of seeing a coincidence when the wavepackets are partially or completely indistinguishable, depending on experimental setup. When the wavepackets are completely indistinguishable, $\mathcal{V}_{\text{HOM}}$ corresponds to 0.5. However, in an experimental situation the quantity P_C can deviate from ideal value, and hence the HOM interference might be imperfect ($\mathcal{V}_{\text{HOM}} > 0.5$).

When the two wavepackets are completely distinguishable, the probability of seeing an error is 1/2. This can be observed from Fig. 5.2 where only two of the four cases result in both photons going to the same detector. Therefore, the error rate can be related to the HOM visibility as

$$e_{k_A k_B} = \frac{(1 - \mathcal{V}_{\text{HOM}})}{2}. \tag{5.18}$$

For the case where $k_A = k_B$, and the HOM interference is perfect ($\mathcal{V}_{\text{HOM}} = 0.5$), $e_{k_A k_B} = 0.25$ as discussed above. Similarly, when $k_A \neq k_B$, the HOM visibility is $0.5 < \mathcal{V}_{\text{HOM}} < 1$, and the error rate is $0.25 < e_{k_A k_B} < 0.50$.

Using all the tools developed here, I simulate the secret key rates as shown in Fig. 5.5. For the simulation, I set $p_T = 0.9$, $p_F = 0.1$ and $p_\mu = p_v = p_\omega = 0.33$ and the misalignment error in time and phase basis is set to 1%. The rate of state preparation (frame rate) is set as $2.5/d$ GHz, where d is the dimension. The detector efficiencies are assumed to be a function of the detection rate that takes into account the saturation effects as discussed in Chap. 3. Additionally, the mean photon numbers μ, v, ω are set to 0.4, 0.08, and 0.025, respectively. The probability of observing a dark count event in a given frame is set 10^{-7}.

In Fig. 5.5, lines of different color indicate various quantum channel losses (fiber length). The coefficient of loss in the fiber is assumed to be 0.2 dB/km. It is seen that the optimum value of the dimension that maximizes the secret key rate decreases as a function of channel loss, indicating that high-dimensional protocols are useful at lower-to-moderate channel losses.

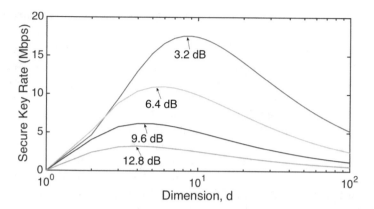

Fig. 5.5 Extractable secret key rates as a function of dimension. Simulated secret key rates as a function of dimension plotted for various quantum channel losses

5.5 Experimental Demonstration

One of the key experimental challenges for this QKD system is to ensure perfect indistinguishability between the incoming photons and the locally generated quantum states. To accomplish this, one needs to ensure that the two sources are indistinguishable in four independent degrees-of-freedom: spectral, polarization, temporal, and spatial [8]. In the experiment, the matching of all four degrees-of-freedom is accomplished as follows.

The temporal overlap is ensured by delaying the quantum states generated in Bob's setup with respect to the incoming quantum states from Alice. The coarse tuning is performed by adjusting the offsets of the FPGA signals that are used to drive the electro-optic modulators; fine-tuning is achieved using a picosecond-resolution optical delay line. The spatial mode overlap is obtained by using single-mode optical fiber throughout the experiment. Matching of the spectral mode is achieved by observing the beatnote frequency of the two sources and tuning it below 10 MHz. Although matching of beatnote frequency with a tunable laser is a reasonable approach for a laboratory-based proof-of-principle experiment, this can be challenging for a field demonstration of the protocol. A better approach is to use two lasers whose frequencies are locked to a molecular absorption line. Finally, the polarization modes of the photons are matched using polarization-maintaining optical fiber throughout the experimental setup and by choosing the correct polarization using polarizing beamsplitters. For a deployed fiber, this can be achieved using a polarization tracker and using active feedback to adjust the polarization of both the incoming photon and the locally generated photon in Bob's setup.

Figure 6.4 shows the experimental setup used to demonstrate this protocol. Since the experiment is performed in a laboratory-setting, the same FPGA is used to generate quantum states in both Alice and Bob's setups. However, the quantum states are generated using two independent continuous-wave lasers. A small fraction of continuous light (2%) from Laser 1 (wavelength tunable Agilent HP81862A) and Laser 2 (Wavelength Reference Clarity NLL-1550-HP locked to a hydrogen cyanide cell) is mixed in a 2 × 1 beam combiner and the beatnote frequency is measured with a high-speed photoreceiver (Miteq DR-125G-A, not shown). The wavelength of Laser 1 is tuned to ensure that the beatnote frequency between the two lasers is under 10 MHz. This ensures spectral overlap of the two sources. The other fractions (98%) from both lasers are intensity modulated using electro-optic intensity modulators (EOSpace) to create the time and phase states.

The cw light from Laser 1 is intensity modulated twice through IM 1 and IM 2 using the same pattern from the FPGA and is temporally overlapped to ensure that the signal to noise ratio of the resulting wavepacket is large. Both IM 1 and IM 2 have extinction ratio ~ 20 dB, which when combined give an extinction ratio close to 40 dB. The cw light from Laser 2 is intensity modulated using IM 3 and IM 4 using a similar technique. To create the signal and decoy states of any arbitrary mean photon numbers, the amplitude of the signals going into IM 2 and IM 4 is adjusted using a technique similar to the one used in Chap. 3. The clock rate of the FPGA is set to 10 GHz, which creates wavepacket peaks of width 100 ps (full-width at half-maximum). The time-bin width of the quantum states is set to 400 ps, making an overall clock rate of 2.5 GHz.

The signals from both the lasers are then attenuated using variable optical attenuators (VOA, Fibertronics). The incoming quantum states in Bob's receiver are split using a 90/10 directional coupler for time (90%) and phase (10%) basis measurements. Bob uses an optical delay line (ODL, General Photonics VDL-001-35-60-SS-FC/APC) to tune the temporal overlap of the wavepackets to a resolution of 1 ps and a total scanning time of 600 ps. The two wavepackets are then passed through polarizing beamsplitters (PBS, Thorlabs) to pick the correct polarization and then interfered at a polarization maintaining beamsplitter (Thorlabs). The outputs of the beamsplitter are then coupled into two superconducting nanowire single-photon detectors (SNSPDs, Quantum Opus) and the events are then time-tagged using a 50-ps-resolution time-to-digital converter (Agilent, Acqiris U1051A) (Fig. 5.6).

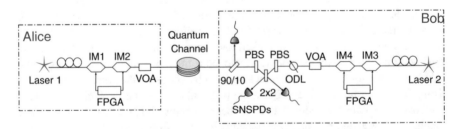

Fig. 5.6 Detailed schematic of the experimental setup. Alice and Bob share the same set of equipment to modulate cw lasers into quantum states with different mean-photon numbers. The quantum states are then overlapped at a beamsplitter to perform a HOM interference

5.5.1 Characterization of HOM Interference

To characterize the quality of indistinguishability, I first look at the HOM interference of two independent laser sources when the spatial, spectral, and polarization degrees-of-freedom are matching but the temporal overlap is varied. Specifically, two single-peaked wavepackets of $\sim 100\,$ps pulse-width from Alice and Bob, generated at a repetition rate of 612.5 MHz with a mean photon number of 0.0014 ± 0.0001, are interfered at a beamsplitter and the coincidence rates are recorded using a time-to-digital converter. The temporal overlap of the wavepackets is varied using an optical delay line and the coincidence rates are recorded at each delay. Figure 5.7 shows the raw coincidence values as a function of the relative time-delay. It is seen that when there is no temporal overlap between the wavepackets, for example at $-200\,$ps and at $200\,$ps, the raw coincidence counts recorded per $4\,$s is 11920 ± 200. The raw coincidence values decrease as the temporal overlap between the wavepacket is increased and a minimum is observed when the two wavepackets are completely overlapping at $0\,$ps. At this point, the raw coincidence events recorded per $4\,$s decreases to 5910 ± 230. The overall drop in coincidence value is 0.50 ± 0.03 which is in good agreement with the theoretical value of 50%.

It is important to emphasize that the data presented here are the raw coincidence counts which is not post-processed to enhance the visibility in any way. Typically, the coincidences due to the background counts are subtracted to eliminate accidental coincidences. In this case, due to the high-extinction ratio of the single-photon wavepacket, low mean photon number, and low dark count rates, no post-processing of the data is necessary. The overall detection rates at the detectors are $(3.0 \pm 0.1) \times 10^5$ counts/second. The low detection rate also reduces the accidental coincidences that occur due to the spurious events of the SNSPDs as discussed in Chap. 3.

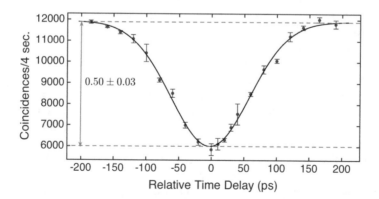

Fig. 5.7 Characterization of the HOM interference visibility. Two single-peaked wavepackets from independent laser sources are interfered at a beamsplitter and the correlated events are recorded using single-photon counting modules. The temporal overlap of the wavepackets are varied using an optical delay line, and the number of coincidence events is recorded at each position

Another crucial point is that the coincidence value does not reduce to zero at the minimum (dip), as expected for perfect HOM interference of single-photon states. This is due to the imperfect nature of the PRWCS used in this experiment, where vacuum and multi-photon components are emitted with finite probability. However, it is possible to extract the visibility due to the single-photon contribution using the decoy-technique, and show that the coincidence rate goes close to zero at the dip. The analysis to extract the single-photon contribution is discussed in the previous section, where the bound for single-photon gain is calculated. For the QKD experiment, the relative delay is set to the value corresponding to the minimum. Hence, the extracted single-photon coincidence rate as a function of time-delay is not presented here; such a plot can be found in Chap. 6 in the context of quantum cloning experiment.

5.5.2   Experimental Parameters

For the proof-of-principle demonstration of the QKD protocol, the FPGA memory is loaded with a fixed pattern consisting of 90% temporal and 10% phase basis states. The ratio of signal to decoy to vacuum is set on the FPGA memory to $1/3 : 1/3 : 1/3$ for both time and phase bases. The mean photon numbers are chosen to be $\mu \sim 0.5$, $\nu \sim 0.11$, and $\omega \sim 0.001$. The time-bin width is set to 400 ps to minimize the error rates while extracting the highest secret key rate.

An important system parameter for this protocol is the time-bin width. There is a trade-off between the time-bin width and the rate at which a secret key can be generated. If the time-bin width is set too small, the quantum bit error rate in the time-basis increases due to the jitter of the single-photon detectors, resolution of the time-to-digital converter, saturation of the single-photon detectors, among others. On the other hand, if the time-bin width is set too large, leakage of light through the electro-optic modulators increases background counts and hence the quantum bit error rate in the time basis increases.

A similar trade-off exists in the phase basis. If the time-bin width is too large, the error rate in phase basis increases as d gets larger due to imperfect overlap of the spectral modes between the two lasers. Specifically, the center frequencies of the two lasers are different, in general (to within 10 MHz of each other), and therefore the phase difference between the two lasers increases as a function of time. If the frequency difference between the two sources is $\Delta\nu$, then the phase difference between the two lasers is $2\pi\Delta\nu T_{frame}$, where T_{frame} is the frame width. To observe a good HOM interference, the two frequencies must be very close to each other, or the time-bin width must be very small, so that the condition $2\pi\Delta\nu T_{frame} \ll \pi$ is satisfied.

In this experiment, it is not possible to adjust the center frequencies of the two lasers < 10 MHz, which means the time-bin width must be optimized to 400 ps, so that the error rates in both the bases are as small as possible. This can be improved in future experiments by using two lasers locked to a molecular absorption line.

5.5.3 System Performance

Incorporating all the experimental and theoretical tools, I implement the protocol and characterize the system based on some key performance metrics. The quantum channel loss is set to 3.2 dB (16 km length in optical fiber of coefficient of loss 0.2 dB/km). Figure 5.8a shows the raw detection rates as a function of dimension for the signal and decoy states in the time basis. For $d = 2$ and $d = 4$, the detection rate of the signal states is very similar, indicating that the detectors are operating in the saturation regime. The raw detection rate decreases as $1/d$ for higher dimensions due to the increasing frame size (decreasing state preparation rate) as can be seen from Fig. 5.8a. In Fig. 5.8b, I plot the coincidence rates for the cases where both Alice and Bob send the same mean photon numbers, $\mu - \mu$ (signal-signal) or $\nu - \nu$ (decoy-decoy). The mean photon numbers μ, ν and ω are set to ~ 0.5, 0.11, and 0.001. They vary slightly from one dimension to the next, but the ratio of the mean photon numbers is fixed. The coincidence rates also decrease as a function of dimension as expected. In this case, the dependence of coincidence rate

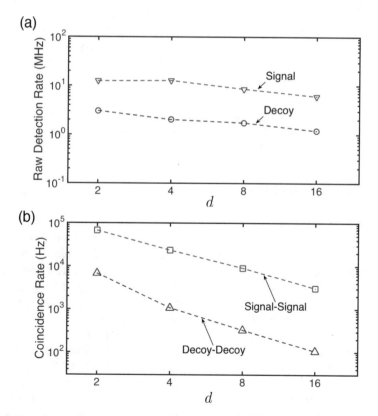

Fig. 5.8 Detection and coincidence rates. The raw detection rates in the time basis (**a**) and the coincidence rates in the phase basis (**b**) are plotted as a function of dimension

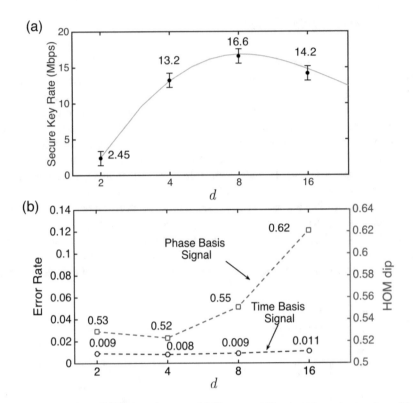

Fig. 5.9 Observation of high secret key rate. (**a**) The extractable secret key rate at a channel loss of 3.2 dB plotted as a function of dimension. (**b**) The corresponding quantum bit error rates in the time basis and the HOM dip values in the phase basis are plotted as a function of dimension for the same channel loss

on the dimension should be $1/d^2$, since coincidence requires a photon from Alice and another from Bob.

Figure 5.9a shows the extractable secret key rate as a function of dimension. The solid line in Fig. 5.9a shows the theoretical secret key rate obtained by matching the parameters in the simulation to the experimentally determined values. The experimental secret key rate is calculated using Eq. (5.10) in which the bounds for $R_{T,1}$, $Y_{F,11}$, $Y_{T,0}$, and $e_{F,11}$ are determined from the experimentally measured detection and error rates. This bound is only moderately tight because I assume $e_{F,11} = e_F^U$, which is only true if all the phase states are transmitted (see Chap. 4). The reason for the assumption is that Matlab cannot run the optimization code which calculates the upper bound of the phase error rate, when the dimension is 16. Recall from Chap. 4 that when $d = 16$, the size of the density matrix is $16^2 \times 16^2$, which means Matlab has to optimize a $16^2 \times 16^2$ phase error matrix and this overloads the memory. Hence, the bound for the secret key rate is only moderately tight, which can be improved in the near future with either an analytic extrapolation of this method or using new computational techniques.

The corresponding quantum bit error rates in the time basis and the HOM dip values in the phase basis are shown in Fig. 5.9b. The HOM dip value is related to HOM visibility as $1 - V_{HOM}$. The time-basis error rate in Fig. 5.9b corresponds to the mean photon number μ, and the HOM dip values correspond to the case where Alice transmits a phase state with mean photon number μ and Bob injects a phase state with mean photon number of μ. The time-basis error rates corresponding to the mean photon number ν are measured to be 0.029 ± 0.002, 0.036 ± 0.001, 0.038 ± 0.002, and 0.059 ± 0.002 for $d = 2, 4, 8$, and 16. The HOM dips for the decoy mean-photon numbers $\nu - \nu$ are determined to be 0.59 ± 0.04, 0.64 ± 0.03, 0.59 ± 0.05, and 0.63 ± 0.04 for $d = 2, 4, 8$, and 16. Finally, to calculate the bound $e_{F,11}$, one needs to estimate the HOM visibility corresponding to the mean photon numbers $\omega - \omega$. The total number of coincidence events recorded for this particular combination of mean photon numbers is between 0–10 for each d. It is not possible to estimate the value of HOM visibility with such low count rates. Therefore, I assume that the HOM visibility corresponding to $\omega - \omega$ is 0.50, which is the theoretical minimum.

From Fig. 5.9, it is seen that the experimental data closely follow the theoretical curve and that the maximum secret key rate is achieved at $d = 8$, which is in agreement with the theory. The corresponding error rates in the time basis is $<$ 1.2%, all the way up to $d = 16$, indicating that the background leakage due to finite-extinction ratio of the intensity modulators is not a major issue in the system. The error rates in the phase basis is obtained from the HOM visibility as $(1 - V_{HOM})/2$. Overall, the error rates in the phase basis are higher than the time basis, and this is likely due to imperfect indistinguishability, mismatch in the polarization and spectral modes, and a slightly unbalanced beamsplitter. Additionally, for higher d values, the frame size increases, and therefore the indistinguishability due to the mismatch of the laser frequencies also increases.

5.6   Conclusion and Future Work

To summarize, in this chapter I demonstrate an arbitrary-dimensional QKD protocol that can be implemented with a simple receiver comprising of a local ancilla source, a beamsplitter, and two single-photon counting modules. I demonstrate a proof-of-principle operation of the system, and simulate the theoretical secret key curves which are in agreement with the experimentally achieved secret key rate.

This experiment paves the way for a flexible and simple scheme that can be used for field implementation of high-dimensional QKD protocols, assuming two frequency-locked lasers are used to generate the quantum states and an active feedback with a polarization tracker is used to adjust the polarization. In future experiments, the secret key rate can be multiplied by a few folds using a wavelength-division multiplexing scheme, where signals encoded at different wavelengths are multiplexed into a single dark fiber. At the receiver, demultiplexers can be placed at the outputs of the beamsplitter and couple into multiple detectors, thereby increasing the secret key rate.

References

1. M. Curty, F. Xu, W. Cui, C.C.W. Lim, K. Tamaki, H.-K. Lo, Nat. Commun. **5**, 1 (2014). https://www.nature.com/articles/ncomms4732
2. N. Gisin, S. Fasel, B. Kraus, H. Zbinden, G. Ribordy, Phys. Rev. A **73**, 022320 (2006). https://doi.org/10.1103/PhysRevA.73.022320
3. J.-Y. Guan, Z. Cao, Y. Liu, G.-L. Shen-Tu, J.S. Pelc, M.M. Fejer, C.-Z. Peng, X. Ma, Q. Zhang, J.-W. Pan, Phys. Rev. Lett. **114**, 180502 (2015). https://doi.org/10.1103/PhysRevLett.114.180502
4. C.K. Hong, Z.Y. Ou, L. Mandel, Phys. Rev. Lett. **59**, 2044 (1987). https://doi.org/10.1103/PhysRevLett.59.2044
5. Y. Liu, T.-Y. Chen, L.-J. Wang, H. Liang, G.-L. Shentu, J. Wang, K. Cui, H.-L. Yin, N.-L. Liu, L. Li, X. Ma, J.S. Pelc, M.M. Fejer, C.-Z. Peng, Q. Zhang, J.-W. Pan, Phys. Rev. Lett. **111**, 130502 (2013). https://doi.org/10.1103/PhysRevLett.111.130502
6. Á. Navarrete, W. Wang, F. Xu, M. Curty, New J. Phys. **20**, 043018 (2018).
7. A. Rubenok, J.A. Slater, P. Chan, I. Lucio-Martinez, W. Tittel, Phys. Rev. Lett. **111**, 130501 (2013). https://doi.org/10.1103/PhysRevLett.111.130501
8. R. Valivarthi, Q. Zhou, C. John, F. Marsili, V.B. Verma, M.D. Shaw, S.W. Nam, D. Oblak, W. Tittel, Quantum Sci. Tech. **2**, 04LT01 (2017). http://stacks.iop.org/2058-9565/2/i=4/a=04LT01
9. F. Xu, M. Curty, B. Qi, H.-K. Lo, New J. Phys. **15**, 113007 (2013). http://stacks.iop.org/1367-2630/15/i=11/a=113007
10. H.-L. Yin, T.-Y. Chen, Z.-W. Yu, H. Liu, L.-X. You, Y.-H. Zhou, S.-J. Chen, Y. Mao, M.-Q. Huang, W.-J. Zhang, H. Chen, M.J. Li, D. Nolan, F. Zhou, X. Jiang, Z. Wang, Q. Zhang, X.-B. Wang, J.-W. Pan, Phys. Rev. Lett. **117**, 190501 (2016). https://doi.org/10.1103/PhysRevLett.117.190501
11. X. Yuan, Z. Zhang, N. Lütkenhaus, X. Ma, Phys. Rev. A **94**, 062305 (2016). https://doi.org/10.1103/PhysRevA.94.062305

Chapter 6
Cloning of High-Dimensional Quantum States

6.1 Introduction

In Chap. 2, I describe a simple intercept-and-resend attack that an eavesdropper can perform on a QKD link. I explain how Alice encoding information in two or more randomly chosen mutually unbiased bases compels Eve to measure the quantum states in incorrect basis, resulting in increased quantum bit error rate in the raw key. In principle, Eve could circumvent detection if she was in possession of a quantum device that could copy an information-carrying photon on a different photon. Such a device would allow her to copy all the quantum states in the channel, one at a time, and then measure the quantum states after Alice and Bob discuss their basis choice. However, as discussed in Chap. 1, there is a fundamental limiting principle, known as the no-cloning theorem, that forbids Eve from cloning unknown quantum states in the quantum channel.

The no-cloning theorem is one of the central results of quantum information science [13], proving that it is impossible to make an exact copy of an unknown quantum state. The theorem guarantees the security of various quantum communication and computation methods against an eavesdropper [9], prevents faster-than-light communication using entangled quantum states [5], and sets a limit on the noise-figure of optical amplifiers [7], among other applications.

Regarding state cloning, as discussed in greater detail below, Bouchard et al. [2] recently demonstrated a probabilistic universal quantum cloning machine (UQCM) for d-dimension single-photon orbital-angular-momentum states, which exploits the spatial degree-of-freedom of the photon. Their machine relies on engineered two-photon quantum states enabled by the Hong-Ou-Mandel (HOM) effect [6], as demonstrated recently by Zhang et al. [16]. A key resource for Bouchard et al.'s UQCM is the use of single-photon states generated by heralded spontaneous down conversion sources. Although the single-photon states are critical for a UQCM, the performance of a UQCM can also be estimated using imperfect sources, such as

© Springer Nature Switzerland AG 2018
N. T. Islam, *High-Rate, High-Dimensional Quantum Key Distribution Systems*,
Springer Theses, https://doi.org/10.1007/978-3-319-98929-7_6

phase randomized weak coherent sources (PRWCS) using decoy-state technique. This is demonstrated in Chaps. 3–5 in the context of QKD experiments.

Here, I show the generality of the decoy-state technique by placing tight bounds on the outcome of a quantum cloning machine, where the high-dimensional quantum states are prepared using PRWCS. Specifically, I use the three-intensity decoy-state technique implemented in the QKD system described in Chap. 5, for bounding the single-photon outcome of the UQCM. This study applies the general formalism developed by Yuan et al. [15] to a concrete and complex example of a UQCM, where my results are consistent with Bouchard et al. This paves the way for adapting Yuan et al.'s formalism for even more complex problems, such as Boson sampling [1], where realization of high-purity multi-photon sources remains a challenge.

This chapter is outlined as follows. First, I introduce the no-cloning theorem and show a simple proof using linearity of quantum operators. I then explain optimal quantum cloning, that is, cloning of unknown quantum states with less than unity success probabilities. I also define the terminologies associated with quantum cloning. Then, I explain how to build a UQCM using linear optics and single-photon sources. I then show that the performance of a UQCM can be estimated using the decoy-state technique developed in Chap. 5. I experimentally implement a modified UQCM in which the single-photon sources are replaced with PRWCS. Finally, using the decoy method similar to the ones discussed in the previous chapters, I estimate the performance of a UQCM, and show that it is near optimal.

6.2   Quantum Cloning

6.2.1   The No-Cloning Theorem

The no-cloning theorem was first introduced in 1982 by Wootters and Zurek [13]. Below, I provide a sketch of their proof using the linearity of quantum operators [8].

Consider a photon in an unknown quantum state $|\psi\rangle$, that is to be cloned, is input in an ideal quantum cloning machine. The initial state of the device is set to $|\mathcal{M}\rangle$. To hold the copy of the input photon, a reference quantum state $|R\rangle$ is prepared, much like the purpose of a blank page in a photocopier machine. Suppose, the quantum cloning operation is defined with a unitary operator U that transforms the input product state $|\psi\rangle|R\rangle|\mathcal{M}\rangle$. The action of an ideal quantum cloning machine on an unknown quantum state $|\psi\rangle$ can be expressed as

$$|\psi\rangle|R\rangle|\mathcal{M}\rangle \rightarrow |\psi\rangle|\psi\rangle|\mathcal{M}(\psi)\rangle. \tag{6.1}$$

Specifically, consider the cases where the input state in the quantum cloner are the two computational basis states $|0\rangle$ and $|1\rangle$. An ideal quantum cloning machine acts on the input quantum states $|0\rangle$ and $|1\rangle$ and generates perfect copies,

$$|0\rangle|R\rangle|M\rangle \xrightarrow{U} |0\rangle|0\rangle|\mathcal{M}(0)\rangle, \qquad (6.2)$$

$$|1\rangle|R\rangle|M\rangle \xrightarrow{U} |1\rangle|1\rangle|\mathcal{M}(1)\rangle. \qquad (6.3)$$

Now consider the action of the ideal quantum cloner on the superposition state $|+\rangle$

$$(|0\rangle + |1\rangle)|R\rangle|\mathcal{M}\rangle \rightarrow (|0\rangle + |1\rangle)(|0\rangle + |1\rangle)|\mathcal{M}(0+1)\rangle,$$

$$\rightarrow (|0\rangle|0\rangle + |0\rangle|1\rangle + |1\rangle|0\rangle + |1\rangle|1\rangle)|\mathcal{M}(0+1)\rangle. \qquad (6.4)$$

Another equivalent way of expressing this transformation is to expand the left-hand side of Eq. (6.4) and apply Eqs. (6.2) and (6.3),

$$(|0\rangle + |1\rangle)|R\rangle|M\rangle = |0\rangle|R\rangle|M\rangle + |1\rangle|R\rangle|M\rangle \rightarrow |0\rangle|0\rangle|M(0)\rangle + |1\rangle|1\rangle|M(1)\rangle. \qquad (6.5)$$

By inspection, it is seen that Eqs. (6.4) and (6.5) are not the same, indicating a contradiction. This means the cloner which is able to clone the quantum states in the computational basis is unable to clone the quantum states in the superposition basis. In general, this applies to any unknown quantum state written as $|\psi\rangle = \alpha|0\rangle + \beta|1\rangle$, where α and β are determined from normalization.

6.2.2 Definitions and Terminologies

The proof above exploits the linearity of the quantum operator U to show that a cloner capable of cloning a set of states in a given basis cannot clone quantum states in a different basis. While a perfect copy cannot be made, a degraded copy is possible [3, 4, 12]. Typically, the imperfect clone is characterized by the fidelity $\mathcal{F}$, which is related to the absolute value of the overlap between the initial and cloned states

$$\mathcal{F} = \langle \psi | \rho_i | \psi \rangle, \qquad (6.6)$$

where ρ_i is the reduced density operator of the final state. The final state consists of the input, reference, and the machine state in a product form $|\psi\rangle|R\rangle|\mathcal{M}\rangle$. The reduced density operator is determined by calculating the density operator of the final state $\rho = (|\psi\rangle|R\rangle|\mathcal{M}\rangle)(\langle\psi|\langle R|\langle\mathcal{M}|)$ and then tracing over the reference and the machine state (Fig. 6.1).

A quantum cloner that can achieve optimal cloning fidelity is known as an optimal cloning machine. A quantum cloner that can clone all input quantum states equally well is known as a universal quantum cloning machine. In other words, a UQCM clones any input states with the same fidelity. In the most general case, a quantum cloner is defined as an $N \rightarrow M$ cloner, which means that a state consisting

Fig. 6.1 A high-level
schematic of an ideal
quantum cloner. An ideal
quantum cloning machine
copies an input quantum state
$|\psi\rangle$ into the blank reference
state $|R\rangle$

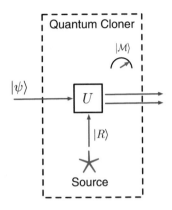

of N input photons is copied into $M - N$ photons. The total number of photons at the
output is M. Typically, experimental implementations of quantum cloning machines
are limited to $N = 1$ and $M = 2$, mainly because it is difficult to prepare and
measure arbitrary photon-number states. The optimal cloning fidelities for a $1 \to 2$
cloner decreases as a function of dimension d, as $\mathcal{F}_{\text{opt}} = 1/2 + 1/(d + 1)$ [4] for
ideally prepared and measured quantum states, demonstrating that high-dimensional
quantum states ($d > 2$) are more robust against cloning compared to qubit states
($d = 2$).

6.3 Universal Optimal Quantum Cloning Machine

Here, I focus on cloning pure single-photon states. Within this context, the machine
is optimal when $\mathcal{F}$ takes on the largest possible value and it is universal when $\mathcal{F}$ is
same for all input states.

Figure 6.2 illustrates the UQCM used by Bouchard et al. A single-photon
quantum state $|\psi\rangle$ is injected to one input port of a symmetric (50:50) beamsplitter
and a reference ancilla quantum photonic state $|R\rangle$ is injected to the other input
port. The ancilla, which mediates the cloning process, makes this device quantum
controlled because a successful cloning of the incoming quantum state depends on
$|R\rangle$. The ancilla photon is a mixed state of dimension d, the density matrix of which
can be represented as I_d/d, where I_d is the identity. In particular, when both $|\psi\rangle$
and $|R\rangle$ are *identical* single-photon states, the HOM effect dictates that the photons
fuse and go in one or the other output ports of the beamsplitter.

The cloned states are created by a second beamsplitter placed in one of the
output ports of the first beamsplitter. With probability 50%, the fused photons split,
resulting in the states $|\Psi_1\rangle$ and $|\Psi_2\rangle$. When this happens, the machine is referred to
as an $1 \to 2$ cloner because one input photon is now copied on to two identical output
photons. An important property of this $1 \to 2$ UQCM is that both the output photons
(copy and original) have the same fidelities, meaning the UQCM is also symmetric.

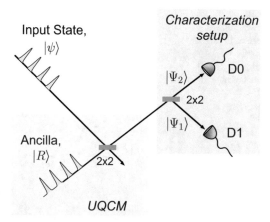

Fig. 6.2 A universal quantum cloning machine. An unknown quantum state $|\psi\rangle$ and a completely mixed ancilla state $|R\rangle$ are input from the two input ports of a beamsplitter. One of the output ports of the first beamsplitter is coupled into the input of a second beamsplitter. The cloning fidelity is measured based on the pattern of detection events registered in the two single-photon counting modules at the outputs of the second beamsplitter

As shown by Bouchard et al., the fidelity of the cloned state can be determined by directing and measuring the states $|\Psi_1\rangle$ and $|\Psi_2\rangle$ to two single-photon counting detectors $D0$ and $D1$, respectively, and recording coincidence counts. For high-fidelity cloning, the coincidence rate observed at detectors $D0$ and $D1$ is high, due to the HOM effect at the first beamsplitter. If the input state $|\Psi\rangle$ and the ancilla state $|R\rangle$ are generated from independent PRWCS, the HOM interference is imperfect. As discussed in Chap. 5, this is due to the probabilistic nature of PRWCS, where some of the states may contain zero or more than one photon.

6.4 Experimental Implementation

Using these experimental and theoretical tools, I implement a UQCM to extract single-photon cloning fidelity from two independent PRWCS. Specifically, I consider the case where Alice transmits quantum states in the superposition basis. These states are prepared by modulating a continuous wave (cw) laser using an electro-optic intensity modulator so that there are d-contiguous peaks each occupying a time-bin. The wavepacket peaks are then phase modulated to prepare states that are orthogonal with respect to each other. Examples of such orthogonal states in $d = 2$ and $d = 4$ are shown in Fig. 6.3a and b, respectively. Although there are other sets of orthogonal states that can be prepared for both $d = 2$ and $d = 4$, this is the simplest set since the phases on the wavepacket peaks are either 0 or π, which can be created with a step function driving a phase modulator.

Fig. 6.3 Input states in $d = 2$ and $d = 4$. Illustration of the superposition states in (a) $d = 2$ and (b) $d = 4$, where the phases on the wavepacket peaks are chosen such that each state is orthogonal to the others

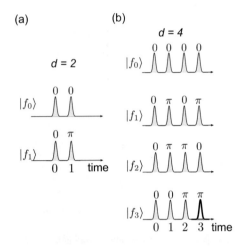

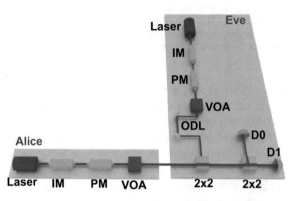

Fig. 6.4 Schematic illustration of the experimental setup. All intensity modulators (IM, EOSpace) and phase modulators (PM, EOSpace) are driven with a field-programmable gate array (FPGA, not shown for clarity). The wavepackets are then attenuated using a variable optical attenuator (VOA) to single-photon level. Eve uses an optical delay line (ODL, General Photonics VDL-001-35-60-SS-FC/APC) to temporally match her wavepackets with the incoming wavepackets from Alice

A schematic illustration of the experimental setup is shown in Fig. 6.4. In this arrangement, Alice and Eve possess identical set of equipment to generate d-dimensional phase states. Specifically, Alice uses an intensity modulator to create the superposition time-bin states from a cw laser with all peaks having an identical phase. The width of the wavepacket is set to $\sim$100 ps and the peaks are separated by an 800 ps time interval. Alice also modulates the phase of individual peaks of the wavepacket to create the distinct phase states as shown in Fig. 6.3.

Similarly, Eve has an independent cw laser source that she modulates to create the peaks and the specific phase values. The quantum states from both Alice and Eve are attenuated using variable optical attenuators to the single-photon level. They then pass through two polarization beamsplitters (not shown) in each station to ensure the polarization of the quantum states is identical. The photons from Alice and Eve then interfere at the first 2×2 beamsplitter, whose one output port is connected to the input of a second 2×2 beamsplitter. The output ports of the second beamsplitter are coupled into two high-efficiency ($> 70\%$) superconducting nano-wire single-photon detectors (SNSPDs, Quantum Opus) that are connected to a high-resolution time-to-digital converter (Agilent, Acqiris U1051A). The time stamps of the coincidence events are then processed on a computer to extract detection statistics.

6.5 Results and Discussion

To characterize the quality of interference, I first consider the cases where Alice and Eve both send the state $|f_0\rangle$. This is the case that results in the HOM interference at the first beamsplitter, and can therefore be used to characterize the quality of interference between the two sources. Typically, a high-visibility HOM interference between two PRWCS requires indistinguishability in four independent degrees-of-freedom: polarization, spectral, temporal, and spatial [11], and matching the power of the two sources.

In this experiment, the polarization degree-of-freedom is matched by using the polarizing beamsplitters. In addition, I use polarization maintaining single-mode fibers, which maintain both the polarization and the spatial mode. The matching of the spectral mode is ensured by observing the beat note frequency of the two lasers using a high-speed photo-receiver (Miteq DR-125G-A), and by tuning the wavelength of one laser (Agilent HP81862A) with respect to the other (Wavelength Reference Clarity NLL-1550-HP) to bring the central frequencies within $10\,\text{MHz}$ of each other. As mentioned in Chap. 5, this requirement can be relaxed if two frequency-locked laser sources are used. Finally, the temporal modes are matched by using an optical delay line in Eve's station to adjust her states with respect to Alice's states.

When Alice and Eve transmit identical states, the coincidence counts in detectors $D0$ and $D1$ as a function of the relative time-delay is expected to increase as the overlap between the two states increases and the sources become more indistinguishable. This is precisely due to the HOM interference effect at the first beamsplitter, which results in bunching of photons at the output ports of the first beamsplitter. When the two states are perfectly distinguishable and single-photon states, the probability that two photons leave from one of the output ports of the first beamsplitter is only 1/2. However, when they are completely indistinguishable, the probability increases to 1, and therefore the coincidences in detectors $D0$ and $D1$ is also expected to increase by 100%.

For weak coherent states, however, the coincidences in the two detectors increases by only 50%. This is due to the imperfect HOM interference visibility of PRWCS as discussed in Chap. 5. This is illustrated in Fig. 6.5, for $d = 2$ (left panel) and $d = 4$ (right panel), where I plot the normalized coincidence values as a function of the relative time-delay between the two sources. I find that the coincidences

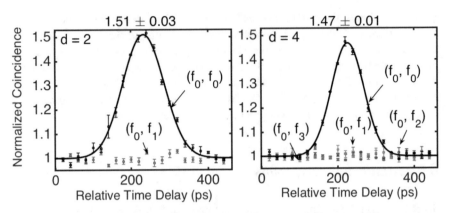

Fig. 6.5 Observation of quantum cloning in $d = 2$ and $d = 4$. The normalized coincidence counts for various input states are plotted as a function of relative time-delay between the two incoming wavepackets for $d = 2$ (left panel) and $d = 4$ (right panel). The black data points represent the case where Alice and Eve transmit $|f_0\rangle$ states. The red, blue, and grey represent the cases where Alice sends $|f_0\rangle$ and Eve transmits $|f_1\rangle$, $|f_2\rangle$ and $|f_3\rangle$

increases by $51 \pm 3\%$ for $d = 2$, and by $47 \pm 1\%$ for $d = 4$. This is in good agreement with the theoretical value of 50%. The small discrepancy between $d = 2$ and $d = 4$ is likely due to the fact that as the dimension increases from 2 to 4, the coincidence window doubles, and therefore the probability of seeing coincidences due to background leakage of the intensity modulator and dark counts also increases. In addition, when the frame size increases, the relative phase difference between the two lasers also increases as discussed in Chap. 5, which results in decreased HOM interference visibility.

For the cases where Alice and Eve transmit orthogonal quantum states, no increase in coincidence is expected because orthogonal quantum states are distinguishable and no HOM interference is observed in the first beamsplitter. This is also illustrated in Fig. 6.5 for $d = 2$ and $d = 4$. The red dots in the left panel indicate the case where Alice sends $|f_0\rangle$ state and Eve sends the orthogonal $|f_1\rangle$. Similarly, the red, blue, and purple data points in the right panel represent the cases where Alice send $|f_0\rangle$ and Eve sends the orthogonal states $|f_1\rangle$, $|f_2\rangle$ and $|f_3\rangle$. For all these cases, it is seen that there is no change in coincidences as a function of relative time delay.

Another way to explain this effect is to realize that a time-dependent phase shift in temporal domain corresponds to a spectral shift in the frequency domain. By adjusting the phase of these states, I am essentially destroying the overlap of the photons in the frequency domain. In other words, for the cases where Alice and Eve send orthogonal states, there are no cloned photons due to the complete distinguishability of the incoming photon states.

The increased coincidence counts observed in Fig. 6.5 represents a successful cloning of the incoming quantum state. The cloning fidelity can be calculated using the visibility of interference [2], or using the coincidence counts of all combinations of input states as

$$\mathcal{F} := \frac{N_{\psi,\psi} + \sum_{i \neq \psi} N_{\psi,i}}{N_{\psi,\psi} + 2\sum_{i \neq \psi} N_{\psi,i}}, \tag{6.7}$$

where $N_{\psi,\psi}$ represent the coincidence rates when the cloning is successful, $N_{\psi,i}$ represent all other cases where the states are orthogonal. The sum of the $N_{\psi,i}$ can be written as $(d-1)\bar{N}_{\psi,i}$, where $\bar{N}_{\psi,i}$ is the averaged coincidence counts for failed cloning. For a d-dimensional system, this simplifies to

$$\mathcal{F} := \frac{1}{2} + \frac{N_{\psi,\psi}/2}{N_{\psi,\psi} + (d-1)\bar{N}_{\psi,i}}, \tag{6.8}$$

$$= \frac{1}{2} + \frac{\mathcal{V}}{d + 2\mathcal{V} - 1}, \tag{6.9}$$

where I define the visibility $\mathcal{V} = N_{\psi,\psi}/(2\bar{N}_{\psi,i})$ between Eqs. (6.8) and (6.9). It is seen that the optimal value is obtained when the normalized coincidence of $N_{\psi,\psi}$ approaches the theoretical value of 2.0. For PRWCS, it only approaches a maximum value of 1.5, which means Eq. (6.8) is not applicable to extract $1 \rightarrow 2$ cloning fidelity when the incoming states are PRWCS.

To place an upper bound on the single-photon cloning fidelity, I implement a two-intensity decoy method. Specifically, I implement the experimental setup such that Alice and Eve can send the quantum states with two different mean photon numbers, $\{\mu, \omega\}$, where $\mu \geq \omega$. I use vacuum ($\omega = 0$, lasers off) as one of the decoy intensities and μ is set to $(3.60 \pm 0.03) \times 10^{-3}$, $(4.90 \pm 0.04) \times 10^{-3}$, $(6.00 \pm 0.10) \times 10^{-3}$, and $(5.70 \pm 0.50) \times 10^{-3}$ for $d = 2, 3, 4$, and 6, respectively. The probability that a coincidence event is detected in the detectors $D0$ and $D1$, given both Alice and Eve have sent a single-photon state is given by

$$Y_{11}^U < \frac{1}{\mu^2}[R_{\mu\mu}e^{2\mu} - (R_{\mu 0} + R_{0\mu})e^{\mu} + R_{00}], \tag{6.10}$$

where R_{ij} is known as the gain of the coincidences when Alice and Eve send states of mean photon numbers i and j. Specifically, the gain is determined by calculating the fraction of the states sent by Alice and Eve that result in coincidence events as discussed in Chap. 5. Essentially, the quantity Y_{11}^U allows to estimate the single-photon probability by measuring coincidence rate when Alice and Eve transmit quantum states of mean photon numbers $\{\mu, 0\}$. The derivation of Eq. (6.10) is shown in Ref. [15].

In greater detail, the calculation of Y_{11}^U is performed as following. First, the raw coincidence counts for each combinations of mean photon numbers $\{\mu, \mu\}$, $\{\mu, 0\}$, and $\{0, \mu\}$ are recorded as a function of temporal delay. Then, the raw coincidence counts are divided by the total number of signals transmitted, which gives an estimate of the gains $R_{\mu\mu}$, $R_{\mu 0}$, and $R_{0\mu}$. The total number of signals transmitted is sum of Alice and Eve's state preparation rate, which for this particular experiment is 2×625 MHz. The gain values and the mean photon number are substituted into Eq. (6.10) to calculate Y_{11}^U. The bound for Y_{11}^U can be converted into 1 photon-1 photon gain by multiplying it with $\mu \exp(\mu) \times \mu \exp(\mu)$, which is the joint probability of Alice transmitting a one-photon state with mean photon number μ and Eve transmitting a one-photon state with mean photon number μ. Finally, it is possible to convert the 1 photon-1 photon gain into the scale of the raw coincidence counts by multiplying it with the sum of Alice and Eve's state preparation rate.

Similarly, a lower bound on the coincidence probability given a single-photon input from Alice and Eve can be obtained with a three-intensity decoy technique, where Alice and Eve transmit one of the three mean photon numbers, $\{\mu, \nu, \omega\}$ [14],

$$
\begin{aligned}
Y_{11}^L \geq \frac{1}{(\mu - \omega)^2 (\nu - \omega)^2 (\mu - \omega)} & [(\mu^2 - \omega^2)(\mu - \omega)(R_{\nu\nu}e^{2\nu} + R_{\omega\omega}e^{2\omega} \\
& - R_{\nu\omega}e^{\nu+\omega} - R_{\omega\nu}e^{\omega+\nu}) - (\nu^2 - \omega^2)(\nu - \omega)(R_{\mu\mu}e^{2\mu} + R_{\omega\omega}e^{2\omega} \\
& - R_{\mu\omega}e^{\mu+\omega} - R_{\omega\mu}e^{\omega+\mu})].
\end{aligned}
\tag{6.11}
$$

Equation (6.11) is identical to that used in Chap. 5. The mean photon numbers for μ and ω remain same as above, and ν is set to $(9.82 \pm 0.09) \times 10^{-4}$, $(1.30 \pm 0.01) \times 10^{-3}$, $(1.60 \pm 0.01) \times 10^{-3}$, and $(1.60 \pm 0.02) \times 10^{-3}$ for $d = 2, 3, 4$, and 6.

Using Eqs. (6.10) and (6.11), I calculate the upper and lower bound of the single-photon gain as a function of relative time delay and normalize it to derive bounds on the coincidence counts as shown in Fig. 6.5. The top panel of Fig. 6.5 shows the raw coincidences for $d = 2, 3, 4$, and 6. The bottom panel shows the upper (red) and lower (blue) bounds of the coincidences calculated using Eqs. (6.10) and (6.11), respectively. It is seen that the upper bound coincidence values increase to the expected value 2 as expected for single-photon sources. In particular, for $d = 2$ and $d = 3$, I obtain good agreement within experimental uncertainties. For $d = 4$ and $d = 6$, the calculated upper bound is slightly lower than expected due to the lower interference visibility at the first beamsplitter. The lower bound of the normalized coincidences calculated using Eq. (6.11) is very close to the upper bound, indicating that the bounds are very tight.

Figure 6.7 shows the cloning fidelity as a function of d extracted from the normalized coincidences in Fig. 6.6 using Eq. (6.9). In these cases, the normalized coincidences for $N_{\psi,i}$ are assumed to be 1, which is a valid assumption based on Fig. 6.5. The blue region indicates the fidelity that can be achieved with weak coherent sources and decoy states. The grey region indicates the fidelity range that is accessible with a coherent attack strategy [10] and the white region indicates the forbidden zone. The bounds for the coherent attack strategies were calculated in Chaps. 3, 4, and 5 for the d-dimensional QKD systems. They correspond to the fidelity of the quantum states after an eavesdropper has interacted with the quantum states using a coherent attack strategy. The red data points show the upper bound on the fidelity and the blue data points show the lower bounds. It is seen that the fidelities are very close to the optimal values that can be achieved with the perfect

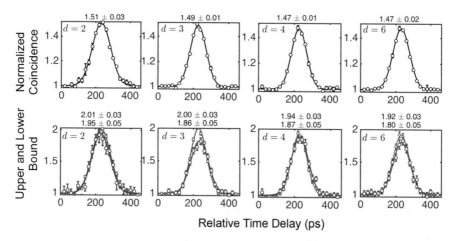

Fig. 6.6 Decoy-state bounds for single-photon quantum cloning. The raw normalized coincidences (top panel) and the normalized upper and lower bounds calculated using the decoy method (bottom panel) as a function of relative time delay for $d = 2, 3, 4$, and 6 (left to right). The red data points show the upper bound extracted using Eq. (6.10) and the blue data points show the lower bound extracted using Eq. (6.11)

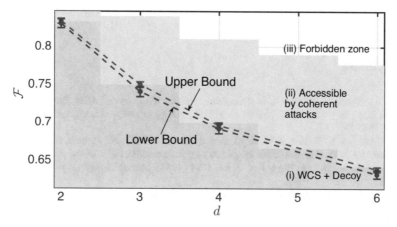

Fig. 6.7 Cloning fidelity plotted as a function of dimension. The blue shaded area indicates the region of fidelity that can be achieved with either a spontaneous parametric down conversion source or weak coherent states with decoy states. The grey area indicates the region that is accessible with coherent attacks. The white area is the forbidden zone that is not accessible with any attack strategies

UQCM. Specifically, the reduced-χ^2 value between the lower (upper) bound data and the optimal fidelities are determined to be 2.68 (0.953). It is also observed that the fidelities are also in agreement with the ones calculated in Ref. [2] where they use high-dimensional OAM states (Fig. 6.7).

The decreasing cloning fidelity as a function of dimension is inherently due to high-dimensional encoding. As d increases, the probability of cloning decreases because the interaction with the d-dimensional ancilla quantum state can now project the incoming state $|\Psi\rangle$ into d possible states. When d is small, the number of possible outcomes is smaller, and therefore there is a greater chance of successfully cloning a quantum state. This is also the reason why high-dimensional QKD protocols can tolerate higher error tolerance than qubit-based protocols.

6.6 Conclusion

To summarize, I present a two-photon quantum cloning experiment, where I use two independent weak coherent sources to extract the cloning fidelity of single-photon states. I rely on the generalized decoy-state technique for two-photon interference and apply it to find both the lower and upper bound of the single-photon cloning fidelity. Finally, I show that the extracted single-photon cloning fidelity decreases as a function of dimension, thereby providing an experimental evidence of why high-dimensional QKD systems can tolerate higher error in the quantum channel than most qubit-based protocols.

References

1. S. Aaronson, A. Arkhipov, in *Proceedings of the 43rd Annual ACM Symposium Theory of Computing*, STOC '11 (ACM, New York, 2011), pp. 333–342. https://doi.org/10.1145/1993636.1993682
2. F. Bouchard, R. Fickler, R.W. Boyd, E. Karimi, Sci. Adv. **3** (2017). https://doi.org/10.1126/sciadv.1601915, http://advances.sciencemag.org/content/3/2/e1601915.full.pdf
3. V. Bužek, M. Hillery, Phys. Rev. A **54**, 1844 (1996). https://doi.org/10.1103/PhysRevA.54.1844
4. N. Gisin, S. Popescu, Phys. Rev. Lett. **83**, 432 (1999). https://doi.org/10.1103/PhysRevLett.83.432
5. N. Herbert, Found. Phys. **12**, 1171 (1982). https://doi.org/10.1007/BF00729622
6. C.K. Hong, Z.Y. Ou, L. Mandel, Phys. Rev. Lett. **59**, 2044 (1987). https://doi.org/10.1103/PhysRevLett.59.2044
7. N. Sangouard, C. Simon, H. de Riedmatten, N. Gisin, Rev. Mod. Phys. **83**, 33 (2011). https://doi.org/10.1103/RevModPhys.83.33
8. V. Scarani, S. Iblisdir, N. Gisin, A. Acín, Rev. Mod. Phys. **77**, 1225 (2005). https://doi.org/10.1103/RevModPhys.77.1225
9. V. Scarani, H. Bechmann-Pasquinucci, N.J. Cerf, M. Dušek, N. Lütkenhaus, M. Peev, Rev. Mod. Phys. **81**, 1301 (2009). https://doi.org/10.1103/RevModPhys.81.1301
10. L. Sheridan, V. Scarani, Phys. Rev. A **82**, 030301 (2010). https://doi.org/10.1103/PhysRevA.82.030301
11. R. Valivarthi, Q. Zhou, C. John, F. Marsili, V.B. Verma, M.D. Shaw, S.W. Nam, D. Oblak, W. Tittel, Quantum Sci. Tech. **2**, 04LT01 (2017). http://stacks.iop.org/2058-9565/2/i=4/a=04LT01
12. R.F. Werner, Phys. Rev. A **58**, 1827 (1998). https://doi.org/10.1103/PhysRevA.58.1827

13. W.K. Wootters, W.H. Zurek, Nature **299**, 802 (1982)
14. F. Xu, M. Curty, B. Qi, H.-K. Lo, New J. Phys. **15**, 113007 (2013). http://stacks.iop.org/1367-2630/15/i=11/a=113007
15. X. Yuan, Z. Zhang, N. Lütkenhaus, X. Ma, Phys. Rev. A **94**, 062305 (2016). https://doi.org/10.1103/PhysRevA.94.062305
16. Y. Zhang, F.S. Roux, T. Konrad, M. Agnew, J. Leach, A. Forbes, Sci. Adv. **2** (2016). https://doi.org/10.1126/sciadv.1501165, http://advances.sciencemag.org/content/2/2/e1501165.full.pdf

Chapter 7
Conclusions and Future Experiments

In this thesis, I describe and experimentally demonstrate various QKD protocols based on time-bin qudits that can generate a secret key at high rates, mainly relying on the greater information content of high-dimensional quantum photonic states and multiplexed single-photon detection schemes.

In this chapter, I provide a summary all major accomplishments achieved in this thesis and propose some future experiments that can be performed using the theoretical and experimental tools developed here.

7.1 Summary

In Chap. 1, I discuss the need for a high-speed QKD system in the forthcoming era of quantum computers. I emphasize that, even though some vendors have developed QKD systems for commercial purposes, the rate at which they operate is significantly lower than the current digital communication rates. Hence, the need for high-speed QKD systems with provable security is paramount.

In Chap. 2, I describe a $d = 2$ time-phase QKD scheme. I emphasize that the security of a QKD protocol depends largely on the ability to generate and detect quantum states in two or more mutually unbiased bases. In particular, I describe an intercept-and-resend attack to give an intuitive explanation for the need of mutually unbiased bases, and thus the security of QKD. Furthermore, I define and describe the level of eavesdropping attacks—independent, collective, and coherent—against which security of various protocols are defined. I conclude Chap. 2 with a discussion of how to build a reconfigurable and efficient QKD transmitter, and a receiver using a combination of commercially available and custom-made equipment.

In Chap. 3, I provide the first original work of this thesis, where I extend the idea of $d = 2$ time-phase QKD to implement a $d = 4$ time-phase QKD system. Specifically, I describe an efficient and reconfigurable QKD transmitter and a novel

© Springer Nature Switzerland AG 2018
N. T. Islam, *High-Rate, High-Dimensional Quantum Key Distribution Systems*,
Springer Theses, https://doi.org/10.1007/978-3-319-98929-7_7

high-dimensional phase state detection system consisting of a tree of time-delay interferometers. Additionally, I describe a multiplexed detection scheme to measure the time-basis states consisting of a 1×4 coupler and 4 single-photon counting detectors, which combined with the greater information content of $d = 4$ states, enable the system to achieve record-setting secret key rates of 26.2, 11.9, 7.71, 3.40, and 1.07 Mbps with channel losses of 4, 8, 10, 14, and 16.6 dB, respectively, corresponding to transmission distances of 20, 40, 50, 70, and 83 km in standard telecommunication optical fiber. Furthermore, with the help of Dr. Charles Ci Wen Lim, I present a provable security proof that accounts for generalized (coherent) attacks, the finite length of the exchanged key, and a broad class of experimental imperfections. Finally, using the upper bounds of phase error rate, I simulate the system and demonstrate a good agreement between the experimental data and simulation results.

In Chap. 4, I study the possibility of simplifying the QKD system presented in Chap. 3 and show that it can be simplified significantly while maintaining the provable security. Specifically, I show that the number of monitoring basis states required to secure a QKD system can be reduced from d to just 1, although at the cost of reduced error tolerance. I simulate the secret key rate using a convex optimization technique known as the semi-definite programming. Using this theoretical result, I show that if the experimental quantum-bit error rate is state-dependent, then sometimes reducing the number of phase basis states can generate secret keys at rates similar to the full setup. Finally, I apply these results to two different QKD schemes, the $d = 4$ time-phase QKD scheme discussed in Chap. 3, as well as to a $d = 7$ spatial-modes based scheme.

In Chap. 5, I explore the possibility of simplifying the interferometric-based QKD receiver to just one beamsplitter and two single-photon counting modules. I show that using a local ancilla source at Bob's setup, it is indeed possible to perform the phase basis measurement using a Hong-Ou-Mandel-type interference with the incoming photon. I then implement and present results of this protocol. This work relies on the efficient scheme of Chap. 4, and is applicable for any arbitrary dimensional QKD systems.

In Chap. 6, I study the possibility of an eavesdropper attacking a high-dimensional QKD system using an optimal quantum cloning machine. I use a similar experimental setup and theoretical tools developed in Chap. 5 to demonstrate that the performance of Eve's universal quantum cloning machine can be simulated using coherent states of varying mean photon numbers. I show that in general the cloning fidelity of single-photon states decreases as a function of dimension, thus illustrating that high-dimensional quantum states are difficult to clone. This also provides an intuitive explanation of why high-dimensional QKD protocols can tolerate more error in the quantum channel compared to a $d = 2$ QKD protocol.

7.2 Future Directions

The experimental and theoretical tools developed in this thesis can be used to investigate various interesting physics problems in quantum information science. Below, I present some of these open questions that can be addressed based on the results described in this thesis.

Perhaps the most important question to address at this point is: given the state-of-the-art transmitter and detector technologies, is it possible to increase the secret key rate of QKD schemes by an order-of-magnitude at short to moderate distances? Such an increment will enable many interesting applications, such as one-time pad encrypted video streaming or hosting an encrypted conference call. To increase the secret key rate of QKD systems, one can leverage on the vast amount of work that has been done in the last two decades to increase the bandwidth and rate of digital communication systems. In the most simplest case, a dense-wavelength division multiplexing (WDM) technique can be adopted, where independent QKD transmitters operating at different wavelengths are multiplexed into a dark fiber [1]. The independent transmitters can also be fabricated on nanophotonic platforms, where many different channels can be placed on one monolithic chip [2, 3]. This will readily increase the rate of communication by several factors. In addition, if each QKD link is based on high-dimensional time-bin encoding, then the long recovery time of the single-photon counting modules can be overcome by increasing the dimension of the encoding qudits.

Although WDM requires an independent transmitter for each channel, the number of time-delay interferometers used to detect the phase states remains the same. The delay interferometers operate over a wide range of telecommunication wavelength (1525–1575 nm). As a result, the same set of time-delay interferometers can be used for all wavelengths. This means, the protocol scales well when multiplexed using WDM scheme, and that the rate of communication can be increased by using a large number of WDM channels.

It is also possible to extend the WDM multiplexing technique to implement the quantum controlled QKD protocol described in Chap. 5. The only challenge is to lock the frequencies of all the lasers at the source and at the receiver. This can be achieved by locking the frequencies to molecular absorption lines, and then matching these center frequencies to the independent WDM channels. This will eliminate the need to stabilize the time-delay interferometers.

Another interesting future study would be to extend the protocols described in this thesis from a prepare-and-measure scheme to a measurement-device-independent scheme, where the receiver is untrusted. Such protocols have the advantage that they eliminate all forms of side-channel attacks on the measurement device since the measurement is performed by an untrusted party. Recent studies have successfully validated the protocols for $d = 2$ time-phase states [4–6]. However extending this to beyond $d = 2$ remains a challenge. The difficulty arises from the fact that the untrusted third party needs to perform a time-reversal Bell-state measurement, which can be easily performed using a beamsplitter and two detectors for $d = 2$ quantum states. However, a beamsplitter is not enough to successfully discriminate $d = 4$ Bell states, and therefore the outcomes to the states input are not one-to-one. This can likely be overcome using more linear optics components.

Besides QKD, there are various interesting quantum optics experiments that can be performed using weak coherent states and the decoy-state technique described in this thesis. Until very recently, it was believed that phase randomized weak coherent states are classical in nature, and therefore cannot be used to perform experiments that require true single-photon sources. However, with the quantum cloning experiment and the generalized decoy frame work by Yuan *et al.* presented in Chap. 6, I have demonstrated that the performance of the true-single photon sources can be simulated using the generalized decoy framework. This opens up the possibilities of using this method for other quantum optics experiments.

There are several possible directions for using this technique in other quantum experiments. One possibility is to implement an $N \rightarrow M$ optimal cloner, where N input photons are cloned to M output photons using weak coherent states [7]. Using several different intensities to generate the quantum states, and a tight bound for each photon number, it might be possible to show that the optimal fidelity is $\mathcal{F}_{N \rightarrow M,opt} = N/M + (M - N)(N + 1)/(M(N + d))$ as proposed in Ref. [8]. Another possibility is to use the optimally cloned quantum states to demonstrate the complementarity of high-dimensional mutually unbiased bases using weak coherent states [9].

References

1. W. Sun, L.-J. Wang, X.-X. Sun, H.-L. Yin, B.-X. Wang, T.-Y. Chen, J.-W. Pan, Integration of quantum key distribution and gigabit-capable passive optical network based on wavelength-division multiplexing (2016). arXiv:1604.07578
2. P. Sibson, C. Erven, M. Godfrey, S. Miki, T. Yamashita, M. Fujiwara, M. Sasaki, H. Terai, M.G. Tanner, C.M. Natarajan et al., Nat. Commun. **8**, 13984 (2017)
3. P. Sibson, J.E. Kennard, S. Stanisic, C. Erven, J.L. O'Brien, M.G. Thompson, Optica **4**, 172 (2017). http://dx.doi.org/10.1364/OPTICA.4.000172
4. H.-L. Yin, T.-Y. Chen, Z.-W. Yu, H. Liu, L.-X. You, Y.-H. Zhou, S.-J. Chen, Y. Mao, M.-Q. Huang, W.-J. Zhang, H. Chen, M.J. Li, D. Nolan, F. Zhou, X. Jiang, Z. Wang, Q. Zhang, X.-B. Wang, J.-W. Pan, Phys. Rev. Lett. **117**, 190501 (2016). http://dx.doi.org/10.1103/PhysRevLett. 117.190501
5. A. Rubenok, J.A. Slater, P. Chan, I. Lucio-Martinez, W. Tittel, Phys. Rev. Lett. **111**, 130501 (2013). http://dx.doi.org/10.1103/PhysRevLett.111.130501
6. R. Valivarthi, Q. Zhou, C. John, F. Marsili, V.B. Verma, M.D. Shaw, S.W. Nam, D. Oblak, W. Tittel, Quantum Sci. Tech. **2**, 04LT01 (2017). http://stacks.iop.org/2058-9565/2/i=4/a= 04LT01
7. E. Nagali, D. Giovannini, L. Marrucci, S. Slussarenko, E. Santamato, F. Sciarrino, Phys. Rev. Lett. **105**, 073602 (2010). http://dx.doi.org/10.1103/PhysRevLett.105.073602
8. M. Keyl and R.F. Werner, J. Math. Phys **40**, 3283 (1999). http://dx.doi.org/10.1063/1.532887
9. G.S. Thekkadath, R.Y. Saaltink, L. Giner, J.S. Lundeen, Phys. Rev. Lett. **119**, 050405 (2017). http://dx.doi.org/10.1103/PhysRevLett.119.050405

Appendix A
Efficiency of the Interferometric Setup

In this appendix, I graphically illustrate that the efficiency of measuring a d-dimensional phase state using an interferometric setup is $1/d$.

Consider the detection of a $d = 2$ phase state using a time-delay interferometer as shown in Fig. A.1. The wavepacket consists of two peaks, each with an electric field amplitude E_0. At the first beamsplitter, the wavepacket is split into two equal parts and half of the wavepacket propagates through the shorter arm of the interferometer, and the other half propagates through the longer arm. The electric field amplitude of each wavepacket peak in each arm is $E_0/\sqrt{2}$, and the intensity (power) is $|E_0|^2/2$. At the second beamsplitter, the wavepackets from both the arms interfere, and a constructive (destructive) interference is observed in the central time bin at the positive (negative) output port of the beamsplitter. The wavepackets occupying the outer time bins do not interfere, and hence are regarded as inconclusive events. Since a detection is only valid if it arrives in the central time bin of the constructive interference port, the electric field (intensity) of the valid events is E_0 ($|E_0|^2$). The ratio of this to the input power is $1/2$, which is equal to $1/d$.

A similar graphical illustration for $d = 4$ is shown in Fig. A.2, where the electric field amplitude (intensity) in the central time-bin is also E_0 ($|E_0|^2$). Since the total input intensity is $4|E_0|^2$, the fraction of events that result in conclusive outcomes is determined to be $1/4$, and therefore the interferometric setup has an efficiency of only 25%. The same method can be extended to any dimension and it can be shown that for any arbitrary dimensional phase state, the efficiency of the phase state measurement scheme is $1/d$.

© Springer Nature Switzerland AG 2018
N. T. Islam, *High-Rate, High-Dimensional Quantum Key Distribution Systems*,
Springer Theses, https://doi.org/10.1007/978-3-319-98929-7

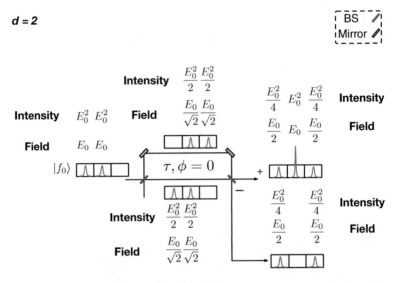

Fig. A.1 Illustration of 1/2 efficiency for a $d = 2$ phase state measurement using interferometric setup. The electric field amplitudes and intensities of the wavepacket as it propagates through different stages of a time-delay interferometer are shown

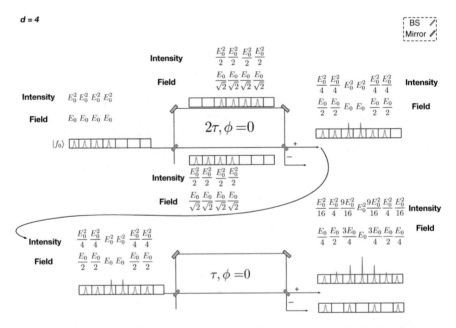

Fig. A.2 Illustration of 1/4 efficiency for a $d = 4$ phase state measurement using interferometric setup. The electric-field amplitudes and intensities of the wavepacket peaks as they propagate through different path of the interferometric setup are shown

Appendix B
Interferometer Characterization

In this appendix, I characterize the stability of commercially available (manufactured by Kylia), passively stabilized, unequal-path-length interferometers, and assess their feasibility for detecting high-dimensional phase states.

B.1 Interferometer Performance

Here, I investigate the performance of two of the Kylia interferometers that are used in the QKD experiment, one with an FSR of 1.25 GHz corresponding to a time-delay of 800 ps, and the other with an FSR of 2.5 GHz corresponding to a time-delay of 400 ps. To characterize the interferometers, I inject a single-frequency, cw laser beam of fixed power into the input port of the interferometer and observe the variation of power from one or both the output ports as shown in Fig. B.1a. The performance of the interferometers is characterized by three primary criteria. First, I measure the long-term ($\sim$ an hour) stability in a nominally controlled laboratory environment where the temperature fluctuates within $\pm 0.1\,°C$. Second, I measure the visibility of the devices over an hour timescale in similar laboratory conditions. Third, I purposely heat the interferometers between 20 and $50\,°C$, and measure the temperature dependent phase-shift (TDPS) of the devices.

The specification of the manufacturer suggests that the TDPS of these interferometers is expected to be <50% of the FSR over a temperature range of 0–$70\,°C$. Based on this specification, I expect that the resonance frequency of the interferometers to shift by less than 10 MHz for a temperature variation of $T < 0.5\,°C$. To ensure that I can observe the small drift in the interferometers phase, I use a frequency-stabilized laser (Wavelength Reference Clarity-NLL-1550-HP locked to an HCN line and operating in the 'Line Narrowing' mode). The laser is specified to have an absolute accuracy of $\leq \pm 0.3\,pm$, and a long-term root-mean-square (RMS) frequency stability better than 1 MHz.

© Springer Nature Switzerland AG 2018
N. T. Islam, *High-Rate, High-Dimensional Quantum Key Distribution Systems*,
Springer Theses, https://doi.org/10.1007/978-3-319-98929-7

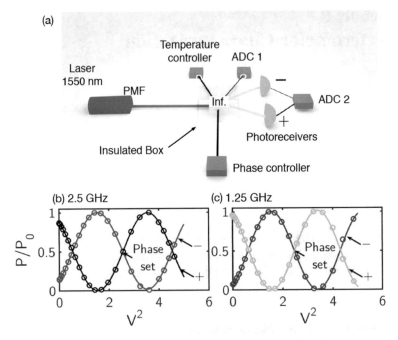

Fig. B.1 Illustration of the experimental setup. (**a**) A cw laser beam (200 μW) is injected into the interferometer using a polarization maintaining fiber (PMF). The temperature of the interferometer is monitored using thermocouples placed at different locations on the aluminum housing of the interferometer. The thermocouple readings are digitized using ADC 1 (National Instruments NI 9239). The output powers are recorded using two photoreceivers (New Focus 2011) and digitized using ADC 2 (National Instruments NI 9239). Interference fringes are observed at the two outputs of the (**b**) 2.5 GHz and (**c**) 1.25 GHz interferometers when the phase ϕ of the interferometers is tuned by applying a voltage to the resistive heater of the interferometers. Adopted from N. T. Islam, A. Aragoneses, A. Lezama, J. Kim, and D. J. Gauthier, 'Robust and stable delay interferometers with application to d-dimensional time-frequency quantum key distribution,' Phys. Rev. Appl. **7**, 044010 (2017), Fig. 4

The characterization measurements of the interferometers are performed by placing the device in a thermally insulated box and allowing it to equilibrate with the environmental temperature for ~2 h with a mean initial temperature of $21.3 \pm 0.3\,°C$. For the stability and TDPS analyses, the phase of the interferometers is initially set to half the fringe maximum, corresponding to the steepest slope ($\phi = \pi/2$) as shown in Fig. B.1b, c. If the phase of the interferometer drifts during the equilibration process, I tune the voltage applied to the heater slightly to set the phase back to the steepest slope. For the visibility measurement, the phase is set to the fringe maximum. Finally, for the TDPS measurement, the temperature of the interferometer is varied between 20 and 50 °C by using heating tapes wrapped around the device. The rate of heating, as well as the final temperature, can be controlled by adjusting the voltage applied to heating tapes.

B.1.1 Stability at Nominally Constant Temperature

The optical power emerging from the two output ports ($\pm$) of an ideal time time-delay interferometer is given by

$$P_{\text{out},\pm} = \frac{\alpha P_0}{2}[1 \pm \cos(k\Delta L)], \qquad (B.1)$$

where P_0 is the power injected at the input of the interferometer, ΔL is the optical path length difference, and $k = 2\pi/\lambda$ is the wave vector. The parameter $\alpha \in \{0, 1\}$ represents the scaling factor that accounts for the reduced transmission due to insertion loss of the interferometer. Assuming that the output power fluctuation comes from the environmental noise which drifts the optical path length by δL, the power at the output port is given by

$$\frac{2P_{\text{out},\pm}}{\alpha P_0} = [1 \pm \cos\{k(\Delta L_0 + \delta L)\}]$$

$$= [1 \pm \cos(\phi \pm k\delta L)], \qquad (B.2)$$

$$= [1 \mp \sin(k\delta L)], \qquad (B.3)$$

where $\phi \equiv k\Delta L_0 = \pi/2$ is used between Eqs. (B.2) and (B.3). Equation (B.3) gives a direct measure of the optical path length drift δL and relates it to the variation of optical power at the output of the interferometer. This measure of the interferometric drift assumes that the injected light into the interferometers has a stable frequency.

The output power of the interferometer can also change due to other effects, which may result in systematic and random errors in the drift measurement. One source of systematic error is the laser power fluctuations. To decouple this effect from the actual interferometric drift, I place a 50/50 beamsplitter after the laser, and monitoring the power variation in the reference $P_r(t)$, while the other fraction is directed to the interferometer. The parameter α in Eq. (B.2) is determined by calculating the ratio of the peak power emitted from the output port of the interferometers and the reference power $P_r(t)$.

Incorporating these corrections for the systematic error, I measure the drift of the interferometers over an hour when the temperature of the environment is stabilized $\pm 0.1\,°C$. I find that both the 2.5 and 1.25 GHz interferometers drift less than 3 nm over 60 min. The variation in δL as a function of time for the 2.5 GHz interferometer is shown Fig. B.2a, and the corresponding change in temperature ΔT as a function of time is shown in Fig. B.2b. In Fig. B.2c, I plot δL as a function of ΔT. I observe that the drift is not completely correlated to the temperature change. The correlation coefficient of 0.8 indicates that there are other factors which could influence the measurement.

There are several factors that could explain the lack of stronger correlation. First, the laser frequency is only stable to a frequency of 1 MHz, which corresponds to a drift of 0.62 nm for the 2.5 GHz interferometer. This is indicated by the error bars in Fig. B.2a, c for the 2.5 GHz interferometers. To estimate the path length drift of

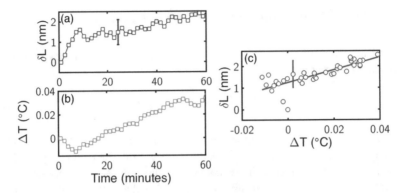

Fig. B.2 Nominally constant temperature drift of 2.5 GHz device. (**a**) The path length drift of the 2.5 GHz interferometer measured over an hour. (**b**) The corresponding temperature variation monitored over the same period of time. (**c**) The path length drift from (**a**) plotted as a function of the temperature variation from (**b**). Adopted from N. T. Islam, A. Aragoneses, A. Lezama, J. Kim, and D. J. Gauthier, 'Robust and stable delay interferometers with application to d-dimensional time-frequency quantum key distribution,' Phys. Rev. Appl. **7**, 044010 (2017), Fig. 5

this interferometer, I fit δL as a function of ΔT in Fig. B.2c with a straight line. From the slope of the fit, I estimate that during the 60 min of data collection, the interferometer drifts a total of 1.2 ± 0.1 nm, which could be partially explained by fluctuation in laser frequency variation. From the same measurement, I also estimate the root-mean-squared error (RMSE) between the fit and data. The measured RMSE of 0.32 nm corresponds to a laser frequency variation of 0.51 MHz, which is well below the specified frequency fluctuation of 1 MHz, and therefore I attribute these small scale fluctuation in the measurement of drift to the laser frequency.

The lack of strong correlation can also be explained by lack of thermal contact between the aluminum housing and the plate on which the optics are mounted. The temperature of the interferometer is measured on the outer aluminum housing, which is not in direct thermal contact with the base plate where the optics are placed. This could potentially lead to a lag in heat transfer, and a low correlation between δL and ΔT.

In Fig. B.3, I show the same plots for the 1.25 GHz interferometer. I observe that during the first 20 min of the run, the interferometer drifts ~ 1.2 nm, and then stabilizes during the rest of the run. As before, there is very little correlation (correlation coefficient of -0.03) between δL and ΔT. In fact, the straight line fit in Fig. B.3c yields a slope of zero, indicating that no path length change occurs over this temperature variation. Furthermore, the RMSE between the fit and data is determined to be 0.34 nm, which corresponds to a 0.27 MHz drift in the laser frequency. This is well within the specified drift for the laser frequency. For a reference, in Fig. B.3a, b, I include error bars that correspond to the 1 MHz frequency fluctuation of the laser frequency. This now corresponds to a drift of 1.24 nm.

I repeated the above measurements ~ 5 times for each interferometers and obtained several independent data sets. For all data sets, I observe a similar pattern

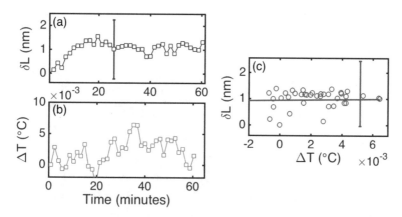

Fig. B.3 Nominally constant temperature drift of 1.25 GHz device. (**a**) The path length drift of the 1.25 GHz interferometer measured over an hour. (**b**) The corresponding temperature variation monitored over the same period of time. (**c**) The path length drift from (**a**) plotted as a function of the temperature variation from (**b**). Adopted from N. T. Islam, A. Aragoneses, A. Lezama, J. Kim, D. J. Gauthier, 'Robust and stable delay interferometers with application to d-dimensional time-frequency quantum key distribution,' Phys. Rev. Appl. **7**, 044010 (2017), Fig. 6

with a lack of correlation between δL and ΔT. Overall, when the temperature of the devices is stabilized to $\pm 0.1\,^\circ$C, the δL is always less than 3 nm. This measurement is an upper bound of the true drift of the interferometers, since the laser frequency fluctuation is comparable to the overall drift.

B.1.2 Visibility

As discussed in the main text related to Eq. (3.5), the interferometer visibility $\mathcal{V}$ is a critical QKD system parameter, which is related to the quantum bit error rate in phase basis. Since phase basis is used to determine the presence of an eavesdropper, visibility represents the amount of mutual information shared between Eve and Alice/Bob. Thus, to extract the largest possible secret key, the base-line change in visibility due to environmental conditions must be characterized and minimized if possible.

To measure the variation in visibility, I inject a cw laser beam into the interferometer (see Fig. B.1a) and measure the output power emerging from the two output ports. For this measurement, the phase of the interferometers is set to 0 rad, which corresponds to the fringe maximum (see Fig. B.1b). The power at the $+(-)$ output port is denoted by $P_{\max}(P_{\min})$, which is same as the probability $\mathcal{P}_+(\mathcal{P}_-)$ defined in Eq. (3.5). For this particular measurement, I do not monitor the reference power from the laser since the typical variation in the laser power ($<0.01\%$) has less than a 0.004% effect on $\mathcal{V}$.

In Fig. B.4, I plot the variation of $\mathcal{V}$ over a time period of 50 min for both the 2.5 GHz (left panel) and the 1.25 GHz interferometer (right panel). The correspond-

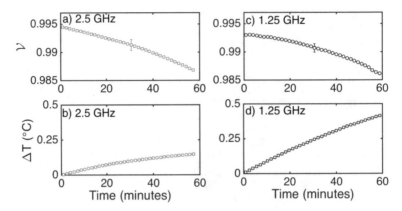

Fig. B.4 Variation of interferometer visibility. The visibility of the 2.5 GHz (**a**) and 1.25 GHz (**c**) interferometer measured over an hour. The corresponding temperature variation of the 2.5 GHz (**b**) and 1.25 GHz (**d**) interferometers measured over an hour. Adopted from N. T. Islam, A. Aragoneses, A. Lezama, J. Kim, and D. J. Gauthier, 'Robust and stable delay interferometers with application to d-dimensional time-frequency quantum key distribution,' Phys. Rev. Appl. **7**, 044010 (2017), Fig. 7

ing temperature fluctuation for the 2.5 and 1.25 GHz devices is shown in Fig. B.4b, d, respectively. Overall, for both the measurements, the temperature changes are slightly larger (±0.5 °C). For both the interferometers, I observe that the visibility remains well above 98.5% during the entire period. The error bars in the plots indicate the change in visibility expected due to the laser fluctuation. These were calculated by propagating the absolute uncertainties through Eq. (3.5), and taking into account the covariance of the dependent variables P_{max} and P_{min}.

B.1.3 Wide-Range Temperature-Dependent Path-Length Shift

To obtain a better estimate of the temperature dependence on the phase shift, that is not as sensitive to the laser frequency fluctuations, here I purposely increase the temperature of the interferometers in large steps. The experimental setup used to characterize TDPS is similar to the one described in Sect. B.1, except the temperature is varied by applying a voltage to the heating tapes wrapped around the interferometer. For each change in temperature, I collect the data for at least 6 h, after which I increase the temperature by another step. Overall, the temperature is varied by 30 °C from an initial temperature of ~22 °C.

Figure B.5a shows the path length drift and Fig. B.5b shows the variation in temperature of the outer aluminum housing for the 2.5 GHz interferometer over four intervals of heating. The vertical dashed lines separate the intervals. At the beginning of every interval, when the interferometer is heated to a new temperature, I observe a rapid initial change in temperature which eventually comes to a thermal

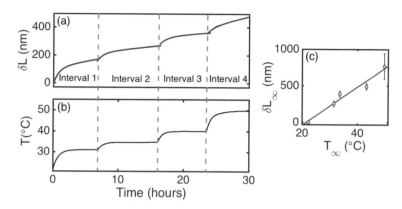

Fig. B.5 TDPS of the 2.5 GHz interferometer. (**a**) The variation in δL and (**b**) temperature for the 2.5 GHz interferometer. (**c**) The extrapolated asymptotic values of δL from each interval plotted as a function of the asymptotic values of T. Red dots represent data and the solid line shows a linear fit. Adopted from N. T. Islam, A. Aragoneses, A. Lezama, J. Kim, and D. J. Gauthier, 'Robust and stable delay interferometers with application to d-dimensional time-frequency quantum key distribution,' Phys. Rev. Appl. **7**, 044010 (2017), Fig. 8

Table B.1 The parameters for the exponential fits on T vs. t

FSR	Intervals	a_0	b_0	c_0
2.5 GHz	1	9.13 ± 0.02	1.223 ± 0.004	21.99 ± 0.02
	2	2.72 ± 0.01	1.370 ± 0.007	32.01 ± 0.02
	3	6.47 ± 0.07	1.48 ± 0.02	33.65 ± 0.07
	4	6.52 ± 0.02	1.047 ± 0.006	43.23 ± 0.02
1.25 GHz	1	4.823 ± 0.008	1.101 ± 0.002	21.97 ± 0.03
	2	6.445 ± 0.001	1.330 ± 0.001	26.92 ± 0.02
	3	18.95 ± 0.05	1.277 ± 0.003	23.67 ± 0.06
	4	7.72 ± 0.03	1.48 ± 0.01	41.75 ± 0.03

equilibrium after $\sim$6 h. Correspondingly, there is a rapid change in δL which slows down over time but keeps increasing monotonically. I find that the temperature change of the outer aluminum housing is well described by an exponential model, $a_0[1 - \exp(-b_0 t)] + c_0$ and the corresponding δL is well described by a double-exponential model, $a_1 \exp(b_1 * t) + a_2 \exp(b_2 * t) + c$, where t is the time and all other variables are parameters obtained from the fitting. The fit parameters for both the models are provided in Tables B.1 and B.2.

Based on these parameters, I observe that the time-scale coefficient, b_1, of the double-exponential represents a characteristic time similar to the time scale of the temperature variation, b_0. This is likely due to the mechanical coupling of the aluminum housing with the glass substrate, which expands and applies stress on the glass substrate. In contrast, the slow coefficient, b_2, represents a much longer time scale, which I assume is due to the eventual expansion of the glass substrate from the temperature change at the beginning of the interval. The weak coupling

Table B.2 The parameters for the double-exponential fits on δL vs. t

FSR	Intervals	a_1	b_1	a_2	b_2	c
2.5 GHz	1	-100 ± 5	-1.4 ± 0.1	-159 ± 34	-0.10 ± 0.04	254 ± 39
	2	-36 ± 3	-1.3 ± 0.1	-100 ± 10	-0.10 ± 0.02	134 ± 13
	3	-41 ± 5	-1.3 ± 0.1	-62 ± 3	-0.23 ± 0.04	101 ± 3
	4	-33 ± 10	-1.6 ± 0.6	-250 ± 158	-0.07 ± 0.07	280 ± 170
1.25 GHz	1	-125 ± 26	-1.4 ± 0.2	-130 ± 55	-0.2 ± 0.1	244 ± 80
	2	-326 ± 127	-1.2 ± 0.2	195 ± 126	-0.6 ± 0.2	113 ± 3
	3	-317 ± 177	-1.3 ± 0.3	311 ± 162	-0.4 ± 0.2	-28 ± 19
	4	-184 ± 9	-1.8 ± 0.1	328 ± 4	-0.24 ± 0.02	-198 ± 12

between the aluminum housing and the glass, as well as the lower heat conductivity of the glass in comparison with aluminum, makes this time scale much longer than each interval period. This is why the plot of δL as a function of time never reaches a stable equilibrium even after 6–7 h of data collection as shown in Fig. B.5b.

To estimate the total change in $\delta L(T)$, I extrapolate the double-exponential (exponential) model to the asymptotic limit. The fit parameters c and c_0 give the asymptotic values for δL and T, which I denote as δL_∞ and T_∞, respectively. These asymptotic values provide an estimate of the drift and temperature that would be attained if the interferometer was allowed to stabilize for a long period of time (longer than $\sim 3/b_2$). I note that this is only an estimate because it assumes a change in δL that is linear with temperature over a long period of time.

Figure B.5c shows a plot δL_∞ as a function of T_∞. To extract the characteristic TDPS of this device, I fit the data with a linear first-order polynomial, from which I determine a slope of 26 ± 9 nm/°C. I define this slope as the TDPS of the 2.5 GHz interferometer.

I also investigate the TDPS characteristic of the 1.25 GHz interferometer. Figure B.6a shows δL and Fig. B.6b shows T as a function of time for this interferometer. By fitting an exponential and a double-exponential model on the temperature and path-length change data, I observe, as before, there are two characteristic time scales for δL. More importantly, I find that the characteristic time scales for these fits are similar to the time scale for the 2.5 GHz interferometer.

However, I also find a sharp contrast in the characteristic δL as a function of temperature for these two interferometers. In Fig. B.6a, during interval 2, as the temperature stabilizes to ~ 33 °C, the corresponding δL changes direction and slopes downward, opposite to the initial change. Similar trend is observed in the subsequent intervals 3 and 4. Specifically, there appears to be a recovery of δL after the rapid initial change due to the temperature variation of the aluminum housing. I predict that this change in direction is due to the expansion of glass, trying to compensate for the change in temperature.

The reversal of δL in intervals 2, 3, and 4 means that I can no longer extract a single value for the TDPS of the interferometer. Indeed, when I extract the asymptotic values of δL and T as before, I see a nonlinearity in the plot of δL_∞ and

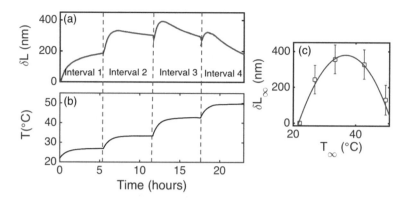

Fig. B.6 TDPS of the 1.25 GHz interferometer. (**a**) The variation in δL and (**b**) temperature variation for the 1.25 GHz interferometer. (**c**) The extrapolated asymptotic values of δL from each interval plotted as a function of the asymptotic values of T. Red dots represent data and the solid line shows a polynomial fit of order 2. Adopted from N. T. Islam, A. Aragoneses, A. Lezama, J. Kim, D. J. Gauthier, 'Robust and stable delay interferometers with application to d-dimensional time-frequency quantum key distribution,' Phys. Rev. Appl. **7**, 044010 (2017), Fig. 9

Table B.3 The key characteristics of the time-delay interferometers

FSR (GHz)	2.5	1.25
Long-term stability (nm)	<3	<3
Change of visibility over an hour (%)	<1	<1
Visibility across 1525 nm and 1565 nm (%)	>99	>99
TDPS (nm/°C)	26 ± 9 over [22, 50] °C	50 ± 17 at 37.1 °C

T_∞. Because of this nonlinearity, it is no longer possible to fit a linear polynomial and obtain a single parameter for the TDPS of the device. Instead, I fit the data with a second-order polynomial. Based on the fit parameters, I extract the slope of the quadratic function at 22 °C to be 50 ± 17 nm/°C and at 37.1 °C to be zero. The latter value represents the temperature at which the plot of δL_∞ vs. T_∞ crosses a local maximum and thus has an instantaneous TDPS of zero.

Although the TDPS of these devices are very small, they still exceed the specifications provided by Kylia for their >2.5 GHz devices (a temperature-dependent frequency shift of less than 50% of the free-spectral range over the 0–70 °C temperature range). The disagreement is most likely due to slight manufacturing errors, such as imperfect thermal compensation, as well as non-ideal effects such as the weak coupling of the substrate with the aluminum housing. Also, the specification only mentions TDPS as the metric of stability, without providing any information on the time-scale over which this measurement is performed. In our investigation of these devices, I find that both the long-term stability at a nominally constant temperature and the TDPS are important metrics of stability for these interferometers. A summary of all our findings from the characterization of these devices is provided in Table B.3.

Appendix C
Numerics-Based Security Analysis

The analysis presented in this appendix is performed in collaboration with Dr. Charles Ci Wen Lim. Specifically, he provided an initial set of Matlab codes for a qubit-based protocol. I extended the analysis to high-dimensional cases based on helpful discussions with him.

C.1 Security Analysis Framework

I analyze the security of the protocol by casting the problem into an optimization framework [1, 2, 5]. The framework, as discussed in Chap. 4, requires all a priori known statistics to maximize the mutual information shared between Alice and Eve.

Analysis of qudit protocols requires promoting Pauli matrices, which are used as basis for qubit-based protocols, to higher dimensions. Specifically, I require a high-dimensional basis that satisfies two specific criteria: (1) Excluding the identity matrix, all other Pauli-equivalent operators in high-dimensional space must be traceless, and (2) the operators must be orthogonal with respect to each other. One set of matrices that satisfies these requirements are Weyl operators [6]. For a d-dimensional system, one can find d^2 of these operators

$$U_{nm} = \sum_{k=0}^{d-1} e^{\frac{2\pi i k n}{d}} |k\rangle \langle k + m|, \quad n, m = 0, 1, \ldots, d - 1, \tag{C.1}$$

where states $|k\rangle$ and $|k + m\rangle$ are in computational basis.

In an equivalent entanglement distillation version, Alice prepares an entangled state of the form

$$|\phi\rangle_{AB} = \frac{1}{\sqrt{d}} \sum_{k=0}^{d-1} |k\rangle_A |k\rangle_B, \tag{C.2}$$

© Springer Nature Switzerland AG 2018
N. T. Islam, *High-Rate, High-Dimensional Quantum Key Distribution Systems*,
Springer Theses, https://doi.org/10.1007/978-3-319-98929-7

where Alice chooses to measure either in $\{T, F\}$ and Bob chooses to measure $\{T, F^*\}$. Note that the random basis choice of $\{T, F\}$ means that Alice performs a trivial unitary operation I for T and non-trivial transformation $\mathcal{H}$ for F before sending the states to Bob. Here, $\mathcal{H}$ represents an operator that performs discrete Fourier transformation. Upon arrival of the signal states, Bob performs a unitary operation I for T and $\mathcal{H}^{-1}$ for F.

I assume that Eve's interaction with Alice and Bob's qudits is independent and identically distributed (i.i.d) such that Eve holds an ancillary on Alice and Bob's qudits. The fact that I assume Eve interacts with Alice and Bob qudits i.i.d means that the state shared among Alice, Bob and Eve can be written as $|\Psi\rangle_{ABE} = \sum_j \sqrt{\lambda_j}|\phi\rangle_{AB}|j\rangle$, such that Eve holds a purification of an ancillary

$$\rho_E = \text{Tr}_{AB}[\Psi\rangle_{ABE}\langle\Psi|]. \tag{C.3}$$

The assumption of i.i.d means that Eve's attacks on Alice and Bob's qudit is collective in nature. It is well known that if the quantum states are permutationally invariant, then it is possible to promote the security for collective attack to general coherent attack using the de Finetti theorem [7, 8]. Note that the quantum states shared between Alice and Bob consists of blocks of time bins or frames which contains vital coherence information. These states are permutationally invariant only if they are considered as frames and not as individual time bins [5].

Eve's interaction with Alice and Bob's qudits introduces an average disturbance $e_{\mathcal{X}}$, where $\mathcal{X} \in \{T, F\}$ represents the basis. The quantum bit error can be written as

$$e_T = \text{Tr}(\rho_{AB} E_T), \tag{C.4}$$

and the phase error can be written as

$$e_F = \text{Tr}(\rho_{AB} E_F), \tag{C.5}$$

where $E_{\mathcal{X}}$ are error operators in the $\mathcal{X} \in \{T, F\}$ basis. The error operator in the T basis and is given by

$$E_T = \sum_{l \in \{0,...,d-1\}} \sum_{k \in \{0,...,d-1\}^*} |l_A, (l+k)_B\rangle\langle l_A, (l+k)_B| \tag{C.6}$$

The asterisk in the index of the sum represents the fact that all the indices should follow the general rule $l + k \neq k$. Similarly, the error operator in the F basis can be written as

$$E_F = (\mathcal{H}_A^\dagger \otimes \mathcal{H}_B)E_T(\mathcal{H}_A \otimes \mathcal{H}_B^\dagger). \tag{C.7}$$

Finally, the projectors can be written as

$$P_l = |l\rangle\langle l| \quad \text{and} \quad P_{\tilde{l}} = |\tilde{l}\rangle\langle\tilde{l}| \tag{C.8}$$

for both the T and F bases.

C.2   Explicit Calculation for $d = 4$

For the temporal basis, T states $|t_0\rangle$, $|t_1\rangle$, $|t_2\rangle$, $|t_3\rangle$, the corresponding phase states
are

$$|f_0\rangle = \frac{1}{2}(|t_0\rangle + |t_1\rangle + |t_2\rangle + |t_3\rangle),$$

$$|f_1\rangle = \frac{1}{2}(|t_0\rangle + i|t_1\rangle - |t_2\rangle - i|t_3\rangle),$$

$$|f_2\rangle = \frac{1}{2}(|t_0\rangle - |t_1\rangle + |t_2\rangle - |t_3\rangle),$$

$$|f_3\rangle = \frac{1}{2}(|t_0\rangle - i|t_1\rangle - |t_2\rangle + i|t_3\rangle). \tag{C.9}$$

Bit error operator in T basis is given by

$$E_T = \begin{pmatrix}
0 & 0 & 0 & 0 & 0 & 0 & 0 & 0 & 0 & 0 & 0 & 0 & 0 & 0 & 0 & 0 \\
0 & 1 & 0 & 0 & 0 & 0 & 0 & 0 & 0 & 0 & 0 & 0 & 0 & 0 & 0 & 0 \\
0 & 0 & 1 & 0 & 0 & 0 & 0 & 0 & 0 & 0 & 0 & 0 & 0 & 0 & 0 & 0 \\
0 & 0 & 0 & 1 & 0 & 0 & 0 & 0 & 0 & 0 & 0 & 0 & 0 & 0 & 0 & 0 \\
0 & 0 & 0 & 0 & 1 & 0 & 0 & 0 & 0 & 0 & 0 & 0 & 0 & 0 & 0 & 0 \\
0 & 0 & 0 & 0 & 0 & 0 & 0 & 0 & 0 & 0 & 0 & 0 & 0 & 0 & 0 & 0 \\
0 & 0 & 0 & 0 & 0 & 0 & 1 & 0 & 0 & 0 & 0 & 0 & 0 & 0 & 0 & 0 \\
0 & 0 & 0 & 0 & 0 & 0 & 0 & 1 & 0 & 0 & 0 & 0 & 0 & 0 & 0 & 0 \\
0 & 0 & 0 & 0 & 0 & 0 & 0 & 0 & 1 & 0 & 0 & 0 & 0 & 0 & 0 & 0 \\
0 & 0 & 0 & 0 & 0 & 0 & 0 & 0 & 0 & 1 & 0 & 0 & 0 & 0 & 0 & 0 \\
0 & 0 & 0 & 0 & 0 & 0 & 0 & 0 & 0 & 0 & 0 & 0 & 0 & 0 & 0 & 0 \\
0 & 0 & 0 & 0 & 0 & 0 & 0 & 0 & 0 & 0 & 0 & 1 & 0 & 0 & 0 & 0 \\
0 & 0 & 0 & 0 & 0 & 0 & 0 & 0 & 0 & 0 & 0 & 0 & 1 & 0 & 0 & 0 \\
0 & 0 & 0 & 0 & 0 & 0 & 0 & 0 & 0 & 0 & 0 & 0 & 0 & 1 & 0 & 0 \\
0 & 0 & 0 & 0 & 0 & 0 & 0 & 0 & 0 & 0 & 0 & 0 & 0 & 0 & 1 & 0 \\
0 & 0 & 0 & 0 & 0 & 0 & 0 & 0 & 0 & 0 & 0 & 0 & 0 & 0 & 0 & 0
\end{pmatrix}. \tag{C.10}$$

The Fourier transform matrix for $d = 4$ is given by

$$\mathcal{H} = \frac{1}{\sqrt{4}} \begin{pmatrix}
1 & 1 & 1 & 1 \\
1 & e^{-\frac{1}{2}(\pi i)} & e^{\frac{1}{2}(-(\pi i))2} & e^{\frac{1}{2}(-(\pi i))3} \\
1 & e^{\frac{1}{2}(-(\pi i))2} & e^{\frac{1}{2}(-(\pi i))4} & e^{\frac{1}{2}(-(\pi i))6} \\
1 & e^{\frac{1}{2}(-(\pi i))3} & e^{\frac{1}{2}(-(\pi i))6} & e^{\frac{1}{2}(-(\pi i))9}
\end{pmatrix} \tag{C.11}$$

Therefore, the error operator for the phase basis can be written as $E_F = (\mathcal{H}_A^\dagger \otimes \mathcal{H}_B) E_T (\mathcal{H}_A \otimes \mathcal{H}_B^\dagger)$. All the joint probabilities where Alice sends in one basis and

Bob measures in the other will be 1/16. All the joint probabilities for partial measurements where Alice sends T and Bob measures in F will be $1/4\,(1-e_T)$ for correct measurements, and $1/12e_T$ for error measurements.

C.3   Matlab Code for the SDP

In experiments, the bit error rate I measure is D_T. The goal of the SDP program is to maximize D_F given the bit error rate and the measurements of the projector operators for any d^2 matrix ρ_{AB}. The only assumptions that go in the optimization is that the trace of ρ_{AB} is unitary and the expectation values of the projection operators. The fact that I am allowing ρ_{AB} to be arbitrary also implies that Eve can perform any arbitrary operations on the states transmitted between Alice and Bob, hence the bound is valid for coherent attacks.

The algorithm for the SDP code is as follows:

```
maximize Tr(E_F ρ_AB) such that,
```
$$\text{Tr}(\rho_{AB}) = 1$$
$$\rho_{AB} \geq 0$$

```
%QBER is the expected experimental error
```
$$\text{Tr}(E_T \rho_{AB}) = e_T$$

```
%All the joint operators where Alice sends F and Bob measures in T
                  %The indices i, j = 1:4
```
$$\text{Tr}(|f_i\rangle\langle f_i| \otimes |t_j\rangle\langle t_j|\rho_{AB}) = \tfrac{1}{16}$$

```
%All the joint operators where Alice sends T and Bob measures in F
                  %The indices i, j = 1:4
```
$$\text{Tr}(|t_i\rangle\langle t_i| \otimes |f_j\rangle\langle f_j|\rho_{AB}) = \tfrac{1}{16}$$

```
%All the partial measurements where Alice sends F and Bob
                   measures in F
```
$$\text{Tr}(|f_0\rangle\langle f_0| \otimes |f_0\rangle\langle f_0|\rho_{AB}) = 1/4 \times (1 - e_T)$$
$$\text{Tr}(|f_0\rangle\langle f_0| \otimes |f_1\rangle\langle f_1|\rho_{AB}) = 1/12 \times e_T$$
$$\text{Tr}(|f_0\rangle\langle f_0| \otimes |f_2\rangle\langle f_2|\rho_{AB}) = 1/12 \times e_T$$
$$\text{Tr}(|f_0\rangle\langle f_0| \otimes |f_3\rangle\langle f_3|\rho_{AB}) = 1/12 \times e_T$$
$$\text{Tr}(|f_1\rangle\langle f_1| \otimes |f_0\rangle\langle f_0|\rho_{AB}) = 1/12 \times e_T$$
$$\text{Tr}(|f_1\rangle\langle f_1| \otimes |f_1\rangle\langle f_1|\rho_{AB}) = 1/4 \times (1 - e_T)$$

$$\text{Tr}(|f_1\rangle\langle f_1| \otimes |f_2\rangle\langle f_2|\rho_{AB}) = 1/12 \times e_T$$
$$\text{Tr}(|f_1\rangle\langle f_1| \otimes |f_3\rangle\langle f_3|\rho_{AB}) = 1/12 \times e_T$$
$$\text{Tr}(|f_2\rangle\langle f_2| \otimes |f_0\rangle\langle f_0|\rho_{AB}) = 1/12 \times e_T$$
$$\text{Tr}(|f_2\rangle\langle f_2| \otimes |f_1\rangle\langle f_1|\rho_{AB}) = 1/12 \times e_T$$
$$\text{Tr}(|f_2\rangle\langle f_2| \otimes |f_2\rangle\langle f_2|\rho_{AB}) = 1/4 \times (1 - e_T)$$
$$\text{Tr}(|f_2\rangle\langle f_2| \otimes |f_3\rangle\langle f_3|\rho_{AB}) = 1/12 \times e_T$$
$$\text{Tr}(|f_3\rangle\langle f_3| \otimes |f_0\rangle\langle f_0|\rho_{AB}) = 1/12 \times e_T$$
$$\text{Tr}(|f_3\rangle\langle f_3| \otimes |f_1\rangle\langle f_1|\rho_{AB}) = 1/12 \times e_T$$
$$\text{Tr}(|f_3\rangle\langle f_3| \otimes |f_2\rangle\langle f_2|\rho_{AB}) = 1/12 \times e_T$$
$$\text{Tr}(|f_3\rangle\langle f_3| \otimes |f_3\rangle\langle f_3|\rho_{AB}) = 1/4 \times (1 - e_T)$$

The last 16 lines show explicitly the partial measurements of only a subset of the F-basis states.

C.4 Decoy State Time-Phase Equations

Here, I combine the calculation above with the decoy-state technique from Ref. [4]. For the sake of completeness, I recreate their result and then explain how to include the phase error rate bound calculated above in the decoy-state method. Note that the calculation here is independent of basis choice, and the same quantities can be calculated for the both time and phase basis.

I model the quantum channel as

$$\eta = \eta_d \eta_{\text{ch}} = \eta_d 10^{-\frac{\alpha l}{10}}, \tag{C.12}$$

where η_d is the efficiency of Bob's detectors and α is the coefficient of loss in fiber, typically assumed to be 0.2 db/km. Assuming independence of the i photon states, the transmittance of a i-photon state detected using a threshold detector can be written as

$$\eta_i = 1 - (1 - \eta)^i. \tag{C.13}$$

The probability that Alice sends an i-photon state and Bob receives a detection event (joint probability) can be written as

$$R_i = Y_i \frac{k^i}{i!} \exp(-k) \tag{C.14}$$

where Y_i is the conditional probability that Bob receives a detection event given Alice sends an i-photon state, and k is the mean photon number. The definition of Y_i takes into account that not all photons transmitted by Alice will arrive at Bob's detectors. Additionally, even if the photons arrive in Bob's receiver, the threshold detectors have an i-dependence probability of detecting it. Hence, I can write this as

$$Y_i = Y_0 + \eta_i - Y_0\eta_i \approx Y_0 + \eta_i, \tag{C.15}$$

where Y_0 is the probability of observing an event due to background event and the approximation assumes $Y_0\eta_i << Y_0 + \eta_i$. The overall gain can be expressed as

$$R_k = \sum_{i=0}^{\infty} R_i = \sum_{i=0}^{\infty} Y_i \frac{k^i}{i!} \exp(-k) \tag{C.16}$$

which is the joint probability that Alice sends a state with a mean photon number k and Bob detects an event. Note the quantities R_k and R_i are in general very different.

Now, I define the overall error rate. First, the error rate for an i-photon state is given by

$$e_i = \frac{e_0 Y_0 + e_d \eta_i}{Y_i} \tag{C.17}$$

where e_d is the misalignment probability, which typically arises from finite-extinction ratio of the intensity modulator, or other optical misalignment. The expression for e_k relates to two situations that can happen when Alice sends an i-photon state. First, because Alice transmits a statistical mixture of coherent states, there will be cases where the states contain zero (vacuum) photons. In such scenario, the expression for e_i assumes that there is a finite probability e_0 with which a background event will be detected in a wrong time-bin, hence an error will occur. Second, there will be an erroneous click in the wrong time-bin from misalignment error. The probability with which this happens is e_d and the probability that a threshold detector detects an event is η_i. Therefore, the overall error rate is given by

$$e_k = \frac{1}{R_k} \sum_{i=0}^{\infty} e_i Y_i \frac{k^i}{i!} \exp(-k). \tag{C.18}$$

Again, note that the quantities e_i and e_k are very different. Simplifying the two expression for R_k and e_k, I get

$$R_k = \sum_{i=0}^{\infty} Y_i \frac{k^i}{i!} \exp(-k)$$

$$= \sum_{i=0}^{\infty} (Y_0 + \eta_i) \frac{k^i}{i!} \exp(-k)$$

$$= \sum_{i=0}^{\infty} \{Y_0 + 1 - (1-\eta)^i\} \frac{k^i}{i!} \exp(-k)$$

$$= \sum_{i=0}^{\infty} \left\{ Y_0 \frac{k^i}{i!} \exp(-k) + \frac{k^i}{i!} \exp(-k) - (1-\eta)^i \frac{k^i}{i!} \exp(-k) \right\}$$

$$= Y_0 + 1 - \exp(-\eta k), \quad \text{and} \tag{C.19}$$

$$e_k = \frac{1}{R_k} \sum_{i=0}^{\infty} e_i Y_i \frac{k^i}{i!} \exp(-k)$$

$$= \frac{1}{R_k} \sum_{i=0}^{\infty} \frac{(e_0 Y_0 + e_d \eta_i)}{Y_i} Y_i \frac{k^i}{i!} \exp(-k)$$

$$= \frac{1}{R_k} \{e_0 Y_0 + e_d[1 - \exp(-\eta k)]\}. \tag{C.20}$$

These are precisely the definitions of the detection rates and error rates used in the simulation throughout this thesis. In Sect. 3.5 these equations, multiplied by the total number of signals sent and probabilities of sending each mean photon number, are used to calculate the sifted events. These are also used in Chaps. 4 and 5 to simulate the detection rates.

C.5 Three-Intensity Decoy-State Technique

Suppose that Alice sends quantum states with three different mean photon numbers, $k \in (\mu, \nu, \omega)$. The signal states μ is assumed to be larger than sum of the two decoy states, ν, ω, i.e., $\nu + \omega < \mu$. In addition, assume that $0 \leq \omega \leq \nu$. First, I have to bound the case where Alice sends out a vacuum state and Bob receives a detection. Using Ref. [4], I calculate $\nu \exp(\omega) R_\omega - \omega \exp(\nu) R_\nu$ as

$$\nu \left[Y_0 + Y_1 \omega + Y_2 \frac{\omega^2}{2!} + Y_3 \frac{\omega^3}{3!} + \dots \right] - \omega \left[Y_0 + Y_1 \nu + Y_2 \frac{\nu^2}{2!} + Y_3 \frac{\nu^3}{3!} + \dots \right]$$

$$= (\nu - \omega) Y_0 - \nu \omega \left[Y_1 - Y_1 + Y_2 \frac{\omega - \nu}{2!} + Y_3 \frac{\omega^2 - \nu^2}{3!} + \dots \right]$$

$$= (\nu - \omega) Y_0 - \nu \omega \sum_{i=2}^{\infty} Y_i \frac{\nu^{i-1} - \omega^{i-1}}{i!}. \tag{C.21}$$

Notice that under the assumption $(0 \leq \omega \leq \nu)$ made previously, the second term in the expression is always positive, and hence a bound on the Y_0 term can be written as

$$Y_0 \geq Y_0^L = \frac{\nu R_\omega \exp(\omega) - \omega R_\nu \exp(\nu)}{\nu - \omega} \tag{C.22}$$

Now, I want to bound the single photon detection rate. I begin with the gain corresponding to the mean photon number μ

$$R_\mu = \sum_{i=0}^{\infty} Y_i \frac{\mu^i \exp(-\mu)}{i!}$$

$$R_\mu e^\mu - Y_0 - Y_1 \mu = \sum_{i=2}^{\infty} Y_i \frac{\mu^i}{i!} \tag{C.23}$$

Calculating $R_\nu \exp(\nu) - R_\omega \exp(\omega)$

$$= Y_0 + Y_1 \nu + \sum_{i=2}^{\infty} Y_i \frac{\nu^i}{i!} - Y_0 - Y_1 \omega - \sum_{i=2}^{\infty} Y_i \frac{\omega^i}{i!}$$

$$= Y_1(\nu - \omega) + \sum_{i=2}^{\infty} Y_i \frac{\nu^i - \omega^i}{i!} \tag{C.24}$$

To bound Eq. (C.24), I use the inequality shown in Ref. [3]

$$\nu^i - \omega^i = \frac{\nu^2 - \omega^2}{\nu + \omega} \sum_{j=0}^{i-1} \nu^{i-1-j} \omega^j$$

$$\leq (\nu^2 - \omega^2)(\nu + \omega)^{i-2} \leq (\nu^2 - \omega^2)\mu^{i-2} \tag{C.25}$$

for $i \geq 2$ and $\nu + \omega < \mu$, and hence the assumption in the beginning of the section. In Eq. (C.25), I also use the inequality

$$\sum_{j=0}^{i-1} \nu^{i-1-j} \omega^j \leq (\nu + \omega)^{i-1} \tag{C.26}$$

for $i \geq 2$. Applying these inequalities, Eq. (C.24) can be bounded as

$$R_\nu \exp(\nu) - R_\omega \exp(\omega) \leq Y_1(\nu - \omega) + \frac{\nu^2 - \omega^2}{\mu^2} \sum_{i=2}^{\infty} Y_i \frac{\mu^i}{i!}$$

$$\leq Y_1(\nu - \omega) + \frac{\nu^2 - \omega^2}{\mu^2} [R_\mu \exp(\mu) - Y_0^L - Y_1 \mu] \tag{C.27}$$

Now, the lower bound on Y_1 can be written as

$$
Y_1 \geq Y_1^L = \frac{\mu}{\mu v - \mu \omega - v^2 + \omega^2} \left[R_v \exp(v) - R_\omega \exp(\omega) \right.
$$

$$
\left. - \frac{v^2 - \omega^2}{\mu^2} (R_\mu \exp(\mu) - Y_0^L) \right] \tag{C.28}
$$

where I use the previously obtained bound Y_0^L. Finally, I want to upper bound the error rate on the single photon. I can expand the overall error rate for v and ω as

$$
e_v R_v \exp(v) = e_0 Y_0 + e_1 v Y_1 + \sum_{i=2}^{\infty} e_i Y_i \frac{v^i}{i!}, \tag{C.29}
$$

$$
e_\omega R_\omega \exp(v) = e_0 Y_0 + e_1 \omega Y_1 + \sum_{i=2}^{\infty} e_i Y_i \frac{\omega^i}{i!} \tag{C.30}
$$

Calculating $e_v R_v \exp(v) - e_\omega R_\omega \exp(v)$, I can upper bound the single-photon error rate as

$$
e_1 \leq e_1^U = \frac{e_v R_v \exp(v) - e_\omega R_\omega \exp(\omega)}{(v - \omega) Y_1^L}. \tag{C.31}
$$

It can be observed from the bound on Y_0 that the maximum occurs when $\omega = 0$ which leads to the optimum combination $\{\mu, v, \omega\} = \{\mu, v, 0\}$.

C.5.1 Simulation

Using all the bounds derived above, it is now possible to calculate the secret key fraction [4]

$$
K \geq R_{T,1} [\log_2 d - H(e_F^U)] - R_T f_{EC} H(e_T) \tag{C.32}
$$

where $\log_2 d$ is the number of bits that can be encoded on a photon; $R_{T,1}$ is the single-photon gain as before, but calculated for the time-basis events; e_F^U is the upper bound on the phase error rate, which is a function of the single-photon error rate in the phase basis $e_{F,1}$ which is calculated above; $e_T := p_\mu e_{T,\mu} + p_v e_{T,v} + p_\omega e_{T,\omega}$ is the observed quantum bit error rate in the time-basis; $R_T := p_\mu R_{T,\mu} + p_v R_{T,v} + p_\omega R_{T,\omega}$ is the weighted gain for all mean photon numbers in the time-basis; f_{EC} is the error-correction efficiency factor taken to be 1.16; $H(x)$ is the d-dimensional Shannon entropy for probability x.

C.5.2   Combining Numerics-Based Analysis with Decoy States

Here, I describe how to use the phase error rate bound derived in Sect. C.1 with the decoy-state technique shown in Sect. C.5. Note that the numerical analysis only provides an upper bound for the phase error rate e_{F}^{U} for a given single-photon quantum bit error rate $e_{\mathsf{F},1}$, which is calculated using the decoy-state technique.

When all d states are transmitted in the phase basis, the upper bound of the phase error rate is equal to the single-photon quantum bit error rate $e_{\mathsf{F}}^{U} = e_{\mathsf{F},1}$, and therefore the numerical analysis is not needed. However, when less than $d - 1$ phase states are transmitted, the upper bound on the phase error rate can be estimated in two steps. First, using Eq. (C.31) estimate the single-photon error rate. Second, using this value for the single-photon quantum bit error rate, an upper bound on the phase error rate can be estimated from Fig. 4.1b. This can be extended to any arbitrary dimension and to any arbitrary number of MUB states.

References

1. P.J. Coles, E.M. Metodiev, N. Lütkenhaus, Nat. Commun. **7** (2016)
2. K.T. Goh, J.-D. Bancal, V. Scarani, New J. Phys. **18**, 045022 (2016)
3. C.C.W. Lim, M. Curty, N. Walenta, F. Xu, H. Zbinden, Phys. Rev. A **89**, 022307 (2014). https://doi.org/10.1103/PhysRevA.89.022307
4. H.-K. Lo, X. Ma, K. Chen, Phys. Rev. Lett. **94**, 230504 (2005). https://doi.org/10.1103/PhysRevLett.94.230504
5. T. Moroder, M. Curty, C.C.W. Lim, L.P. Thinh, H. Zbinden, N. Gisin, Phys. Rev. Lett. **109**, 260501 (2012). https://doi.org/10.1103/PhysRevLett.109.260501
6. G.M. Nikolopoulos, G. Alber, Phys. Rev. A **72**, 032320 (2005). https://doi.org/10.1103/PhysRevA.72.032320
7. R. Renner, Nat. Phys. **3**, 645 (2007). https://www.nature.com/articles/nphys684
8. R. Renner, Int. J. Quantum Inf. **6**, 1 (2008)

Printed in the United States
By Bookmasters